Mathematics and Science

An Adventure in Postage Stamps

William L. Schaaf
Professor Emeritus
Brooklyn College
The City University of New York

NATIONAL COUNCIL OF TEACHERS OF MATHEMATICS
1906 Association Drive, Reston, Virginia 22091

Library of Congress Cataloging in Publication Data:

Schaaf, William Leonard, 1898–
Mathematics and science.

Includes index.
1. Mathematics—History. 2. Science—History. 3. Postage-stamps—Topics—Science. 4. Postage-stamps—Topics—Mathematics. I. Title.
QA21.S35 510′.9 78-1680
ISBN 0-87353-122-1

Printed in the United States of America

The nearer man approaches mathematics
the farther away he moves
from the animals.

STANLEY CASSON
Progress and Catastrophe

. . . perhaps the strangest of all
is the marvel that mathematics
should be possible to a race
akin to the apes.

E. T. BELL
The Development of Mathematics

Contents

List of Illustrations vii

Preface xiii

Prologue 1

1 Priesthood and Populace: 3000–600 B.C. 3

Early Beginnings, 3
Babylonian Mathematics, 5
Egyptian Mathematics, 8

2 Philosophers and Geometers: 600 B.C.–A.D. 400 14

The Rise of Greek Science, 15
The Golden Age, 20
The Decline of Greek Mathematics, 23

3 Physicians and Clerics: A.D. 400–1500 27

A Period of Transition, 27
Mathematics in the Near East, 29
Europe at the Crossroads, 32

4 Mapmakers and Stargazers: 1500–1650 39

Charting Untrod Paths, 39
Scanning the Skies, 46
Astronomical Instruments, 52

5 The Awakening: 1650–1800 56

Natural Philosophers, 57
The Birth of Modern Mathematics, 60
Mathematical Analysis, 67

6 New Horizons: 1800–1900 74

The Liberation of Mathematics, 74
Measurement and Computation, 82
The Age of Electricity, 88

7 The Dawn of a New Century: 1900–1940 94

Pure and Applied Mathematics, 94
The Achievement of Flight, 104
Relativity and Atomic Theory, 107

8 Yesterday and Today: 1940–1975 114

The Nuclear Age, 115
Computers and Automation, 120
The Challenge of Outer Space, 125

Epilogue 135

Appendix A: A Checklist of Stamps—Mathematicians and Scientists 137

Appendix B: A Checklist of Stamps—Applications of Mathematics to Science 144

Index 149

List of Illustrations

The following list identifies each stamp pictured in this publication by its figure number (fig. 1.5, for example, refers to the fifth figure in chapter 1), a brief description, the issuing country, the Scott number, and the page on which the illustration can be found.

Figure		Page
1.1	*Ancient chariot,* Algeria 317	4
1.2	*Cuneiform writing,* Austria B316 *Number concept,* Nicaragua 876	5
1.3	*Hieroglyphic stone carving,* Austria B315 *Rosetta stone,* France 1354	8
1.4	*Imhotep,* Egypt 153	9
1.5	*Finger reckoning,* Iran 1413 *Egyptian scribes,* United Arab Republic 385	10
1.6	*The great pyramids,* Egypt C15	11
2.1	*Ionic capital,* Austria 656 *Greek architecture,* Austria 675	15
2.4	*Pythagorean theorem,* Greece 583, Surinam B189, Greece 584	18
2.5	*Law of Pythagoras,* Nicaragua 748	18
2.7	*Aristotle,* Greece RA91 *Democritus,* Greece 717	19
2.8	*Archimedes' law,* Nicaragua C751 *Archimedes,* Spain 1159	21

Figure		Page
2.9	*Hipparchus,* Greece 835 *Ptolemy,* Yemen Arab Republic (unlisted)	24
2.10	*Roman numerals,* Czechoslovakia 1527, Vatican City 366	25
3.1	*Medieval university,* German Federal Republic 766	28
3.2	*Gutenberg,* German Democratic Republic 1167 *Early printing press,* Finland 242	29
3.3	*al-Kindi,* Iraq 303 *al-Farabi,* Iran 948	30
3.4	*al-Haytham,* Qatar 235, Pakistan 281	31
3.5	*Avicenna,* Qatar 237, Persia B32	31
3.6	*Maimonides,* Grenada 401, Israel 74	32
3.7	*Gerbert, Pope Sylvester II,* France B384, Hungary 511	33
3.8	*Adam Riese,* German Federal Republic 799 *Cusanus,* German Federal Republic 792	34
3.9	*Leonardo da Vinci,* Hungary C109, France 682 *Albrecht Dürer,* Ajman (unlisted), German Democratic Republic 1298	36
4.1	*Mercator projection,* Canada 85 *Ortelius,* Belgium B325 *Mercator,* Belgium B324	41
4.2	*Earth globe,* German Democratic Republic 1403 *Map projections,* Brazil 1298, Turkey 1486, Cuba (unlisted)	42
4.3	*Armillaries,* German Democratic Republic 1406, People's Republic of China 201, Austria 774	44
4.4	*Nautical astrolabe,* Portuguese Guinea 297 *Planisphere,* Iran 1049	45
4.5	*Sextants,* Australian Antarctic Territory L21, Nicaragua C747 *Quadrant,* Turkey B82	46
4.6	*Copernicus,* People's Republic of China 205, French Polynesia C95, German Democratic Republic 1461	47
4.7	*Johannes Kepler,* Austria B282 *Tycho Brahe,* Denmark 300	48
4.8	*Galileo,* USSR 2986, Italy 889	50
4.9	*Venus,* Albania 778 *Saturn,* Albania 782	51

Figure *Page*

4.10 *Telescopes,* USSR 2094, Japan 478, German Democratic Republic 898 52

4.11 *Observatories,* USSR 2092, Korea 94, Japan 591, United States 966 54

4.12 *Radio telescopes,* Israel 496, Great Britain 466, France 1067, Haiti C121 55

5.1 *Simon Stevin,* Belgium B321
Evangelista Torricelli, USSR 2165
Otto von Guericke, German Federal Republic 472 57

5.2 *Magdeburg hemispheres,* German Democratic Republic B150 ... 59

5.3 *Huygens's pendulum clock,* Netherlands B365
Mechanical clock, Austria 865
Christian Huygens, Netherlands B36 59

5.4 *René Descartes,* France 331 61

5.5 *Blaise Pascal,* France B181, France 1038 62

5.6 *Isaac Newton,* France 861, Poland 884, Mexico C377 64

5.7 *Gottfried Wilhelm von Leibniz,* German Federal Republic 962, German Federal Republic 360, Romania 1855 66

5.8 *Leonhard Euler,* German Democratic Republic 353, Switzerland B267, USSR 1932 68

5.10 *Integral sign,* USSR 3151
Joseph Louis Lagrange, France 869
Pierre Simon de Laplace, France B298
Jean d'Alembert, France B332 71

5.11 *Royal Astronomical Society,* Great Britain 616 73

6.1 *Carl Friedrich Gauss,* German Federal Republic 725
Niels Abel, Norway 148 76

6.3 *János Bolyai,* Romania 1345
Farkas Bolyai, Hungary 479
Nikolai Ivanovich Lobachevski, USSR 1575 78

6.4 *William Rowan Hamilton,* Eire 126 80

6.5 *Sonya Kovalevski,* USSR 1570
P. L. Chebyshëv, USSR 1051 81

6.6 *Adolphe Quételet,* Belgium 877
Jean Foucault, France 871 82

Figure		Page
6.7	*Micrometer caliper,* United States 1201 *Outside caliper,* India 392 *Slide micrometer,* Korea 803	83
6.8	*Aerial mapping,* Finland 397 *Transit level,* German Democratic Republic 726 *Theodolite,* Finland 373	84
6.9	*Leveling,* Nigeria 207 *Triangulation,* Tanzania/Uganda/Kenya 228	85
6.10	*Metric system,* Romania 1874, Korea 428, Tanzania/Uganda/Kenya 225, Yugoslavia 1180	86
6.11	*Metric units,* Australia 543, Pakistan 364, Australia 541	87
6.12	*Slide rule,* Romania 1159 *Computation symbols,* Colombia C510	87
6.13	*Early calculating machines,* Denmark 415, German Federal Republic 1123	88
6.14	*Benjamin Franklin,* United States 1073 *Charles Augustin de Coulomb,* France B352 *Alessandro Volta,* Italian Somaliland 99	89
6.15	*Hans Christian Oersted,* Denmark 329 *André Ampère and François Arago,* France 626 *Electromagnet,* Denmark 471	90
6.16	*Boltzmann's law,* Nicaragua C749 *Gustav Robert Kirchhoff,* Germany (Berlin) 9N345	91
6.17	*Maxwell's law,* Nicaragua 880	92
6.18	*Heinrich Hertz,* German Federal Republic 762, German Democratic Republic 354, Czechoslovakia 953	93
7.1	*Bertrand Russell,* India 561 *Henri Poincaré,* France 270 *Srinivasa Ramanujan,* India 369	96
7.2	*Paul Langevin,* France 608 *Loránd Eötvös,* Hungary 840 *Jean Perrin,* France 609	97
7.3	*Wilhelm Roentgen,* German Federal Republic 686, Spain 1460 *Alexander Graham Bell,* Niger Republic C191 *Michael Pupin,* Yugoslavia 594	97
7.4	*Induction motor,* Yugoslavia 448 *Transformer,* Yugoslavia 449 *Electronic control,* Yugoslavia 450 *Nikola Tesla,* Yugoslavia 137, Czechoslovakia 949	99

Figure *Page*

7.5 *Ernest Rutherford,* New Zealand 488, USSR 3888 100

7.6 *Guglielmo Marconi,* French Territory of Afars and Issas C87, Czechoslovakia 952
Marconi spark coil, United States 1500 100

7.7 *Electronic communication,* Portugal 1181
Edouard Branly, Czechoslovakia 951 101

7.8 *Alexander Popov,* Bulgaria 723, USSR 1353, USSR 989 102

7.9 *Edwin Armstrong,* Czechoslovakia 954
De Forest audion, United States C86
Radio, United States 1502
Radio wave, United States 1260
Radio, USSR 3526 103

7.10 *Television,* United States 1501, Libya 348, Italy 648, Yemen Arab Republic (unlisted) 104

7.11 *Aviation pioneers,* Spain C56, Paraguay (unlisted), Dahomey C166, Great Britain 584, United States C45 105

7.12 *Aerodynamics,* Switzerland C45
N. E. Zhukovski, USSR 840, USSR 2774
Sergei Chaplygin, USSR 945
A. F. Mozhaisky, USSR 2772 106

7.13 *The Concorde,* Great Britain 581 107

7.14 *Hendrik Lorentz,* Netherlands B35
J. J. Thomson, Sweden 710 108

7.15 *Max Planck,* Germany (Berlin) 9N92, German Democratic Republic 384
Constant ratio h, German Democratic Republic 383 109

7.16 *Albert Einstein,* Switzerland 549, United States 1285, Poland 882 111

7.17 *Ruggiero Boscovich,* Yugoslavia 595, Croatia 60 112

7.18 *Niels Bohr,* Greenland 58
Atomic structure, Sweden 790, United Nations 60 113

8.1 *Einstein's law,* Nicaragua 878
de Broglie's law, Nicaragua C750
Enrico Fermi, Italy 976
Albert Einstein, Israel 117 116

8.2 *The abhorrence of nuclear warfare,* German Democratic Republic B125, United Nations 227, USSR 2077 117

Figure		Page
8.3	*Peaceful use of atomic energy,* United Nations 133, United States 1070, United States 1200	118
8.4	*Nuclear power plants,* Yugoslavia 584, Hungary 1763, People's Republic of China 392, Pakistan 187	119
8.5	*Electronic computers,* German Democratic Republic 811, German Democratic Republic 1532	120
8.6	*Punched cards and magnetic tape,* Norway 547, Canada 542, Israel 258, Netherlands 451, Australia 531	121
8.7	*Automation,* German Democratic Republic 812, German Democratic Republic 1086, Tunisia 496, Ivory Coast 329	123
8.8	*Applications of computers,* Australia 440, German Democratic Republic 1452, USSR 3081, USSR 3416	124
8.9	*Electronic mechanization,* Albania 1253, German Democratic Republic 902	125
8.10	*Robert H. Goddard,* United States C69 *Robert Esnault-Pelterie,* France 1184	127
8.11	*Konstantin Tsiolkovsky,* USSR 2886, Poland 1178, USSR 1582	127
8.12	*Tsiolkovsky and his rockets,* Nicaragua 879, Bulgaria C92, USSR 1991	128
8.13	*Sputniks,* Romania C49, Romania C51, Poland 875	129
8.14	*Satellites in space,* Colombia C499, Republic of the Congo C27, France 1148, Nigeria 143	131
8.15	*Telecommunication by satellite,* Chile C290, Malagasy Republic 465, Canada 444	132
8.16	*U.S. space pioneers,* United States 1193, United States 1331–32	133
8.17	*Moon landing,* United States 1529, United States 1435, United States 1371, United States C76	134

Preface

Early in my career as a teacher of mathematics I became intrigued by the relation of mathematics and science to our culture, both past and present. I was fascinated by the writings of George Sarton, J. W. N. Sullivan, Lewis Mumford, and Jacob Bronowski, among others.

Some years later my boyhood enthusiasm for stamp collecting was rekindled, and before long the persons and the subjects depicted on the stamps appealed to me far more than the nations that had issued them. I soon came to the conclusion that the postage stamps of the world are, in effect, a mirror of civilization and that multitudes of stamps reflect the impact of mathematics and science on society.

Mathematics is a stern discipline, which many of us are inclined to regard either with awe or with frank dislike. Yet mathematics is one of the oldest intellectual concerns of humanity—second, probably, only to language itself. Like the vernacular, mathematics has been patiently created over thousands of years by countless individuals, most of whom have long since been forgotten. A few have left their names on the scroll of fame, and of these, even fewer have been honored on postage stamps—often by countries other than those of their birth. It is to be regretted that more of the great mathematicians and scientists of the world have not as yet been so recognized.

Mathematics is the language of science and technology. It is an international language, even as the scientific community of today is international. I have tried here to tell as simply as possible the story of mathematics from ancient times to the present and to show, through the medium of stamps, how mathematics and technics have shaped the very fabric of our civilization. If through this book more young people should come to appreciate the vital role of mathematics and science in society by way of philately, one of the world's most popular hobbies, I shall feel amply rewarded.

I take this opportunity to thank the National Council of Teachers of Mathematics, as well as the editors of both *School Science and Mathematics* and the *Journal of Recreational Mathematics,* for permission to make use of previously published material. I am also indebted to the late Carl Boyer for suggesting the idea originally, to Mannis Charosh for valuable counsel and encouragement, and to Morris Kline for his generous help and criticism. Any remaining weaknesses or inaccuracies are my responsibility alone. Last, but not least, I am grateful to my son Peter for the photography and to my wife, Jennie, for her unflagging patience in preparing the manuscript.

—W. L. S.

Boca Raton, Florida

Prologue

PREHISTORIC peoples were nomadic; they wandered over the earth in groups as hunters and herders. The search for food was imperative for survival. Their mathematical attainments were presumably limited to the notion of "large" and "small," "many" or "few"; they could count on their fingers, possibly up to three; their number words often consisted only of "one," "two," "three," "many." As herders, they simply followed their animals wherever they could find pasture. As hunters, they found their direction by using the sun and the stars. Their calendar was based on the phases of the moon, and so they lived by a lunar year of 360 days.

After having wandered for perhaps ten thousand years, primitive peoples took a momentous step, one that was to change their life-style permanently: they began to plant and harvest crops. Henceforth they were to cease wandering and would stay in a given place. In so doing, they gradually established communities. Communal living brought with it greater incentives to progress. It accelerated the development of language, and it encouraged the invention of new tools and utensils. Such demands undoubtedly contributed to the slow evolution of quantitative ideas. Presumably people now began to count by using all ten fingers, thus leading eventually to the universal use of our base-ten system of numeration. The need for building shelters made people more conscious of geometric shape and form. The calendar became more refined as a result of the discovery that the solar calendar had 365 days. These calendar experts and early mathematicians generally were priests and often formed the ruling class in primitive societies.

Priesthood and Populace

3000–600 B.C.

MATHEMATICS has a history of its own, just as science or art has. And like the story of many other human achievements, the story of mathematics has been told on postage stamps. To be sure, a few gaps occur here and there, but we shall fill them in as we go along, at least in part.

EARLY BEGINNINGS

How was mathematics created? By whom? Why? To understand this, we must realize that many inventions were created by a host of unknown persons. For example, who fashioned the first knife? Who first thought of the wheel? And so it was with mathematics. Consider the very numbers that we use. Our system of numeration evolved gradually over many centuries, the fruit of the collective efforts of many unknown people. It is so familiar that we take it for granted. And yet, next to language itself, mathematics is one of the most far-reaching inventions of all time. It was not created by any single individual, nor was it completed in any particular country or in any given century.

In addition to the cumulative influence of the populace, mathematics was also created by a relatively small number of intellectual persons whose unique contributions were the result of their creative minds. Many of these gifted individuals have been honored from time to time by having their portraits on postage stamps. Likewise, some of the achievements of mathematicians—and of the scientists who use the mathematics—are often depicted on commemorative stamps. Just as philately reveals the drama of warfare, the saga of science, the mysteries of religion, or the beauty of the arts, so it reveals the evolution of mathematics and its significance for the human race. As the distinguished English mathematician Godfrey Hardy once observed, the story of mathematics did not begin with Pythagoras and will not end with Einstein.

Our story begins more than five thousand years ago. The very earliest part, before the existence of written records, is lost in the obscurity of time. But there is good reason to believe that mathematics existed even before the art of writing was known.

In any event, the cradle of mathematics is located at the eastern end of the Mediterranean in the general area of Asia Minor. During this period the activities of the Sumerians (who lived north of the Persian Gulf and who were later absorbed by the Babylonians of Mesopotamia), the Assyrians, and the Egyptians must be recognized. Before the details of Babylonian and Egyptian mathematics are related, a few general observations about these ancient peoples may be helpful. Both the Babylonian and the Egyptian cultures were sensuous and mystical. There was unquestionably a very real gap between the priests and learned men on the one hand and the ordinary folk and slaves on the other.

The stamp in figure 1.1 depicts an ancient Egyptian war chariot, with Ramses II (1292–1225 B.C.) battling the Hittites. It is typical of the chariots of Babylonia, Egypt, and Assyria and shows that the wheel was already in common use.

Fig. 1.1. Ancient chariot

The Babylonians were familiar with the periodicity of astronomical events, on which they based an elaborate doctrine of astrology. Babylonian trade fostered the development of weights and measures, as well as money in the form of silver bars. The Egyptians knew less about astronomy but more about medicine than the Babylonians, although Egyptian knowledge of anatomy and surgery was crude and relied heavily on sorcery and the will of the gods. The Assyrians were skilled in weaving and embroidering, as well as in carving and sculpture in jasper, crystal, and basalt. The Egyptians likewise reached a high degree of skill in the arts and handicrafts, in sculpture and architecture, and in pottery and jewelry making. Phoenician sailors improved navigation, and their nation developed a phonetic alphabet. The Babylonians used the base 60 in their measurement of time, vestiges of which are retained today in both time and angle measurement.

BABYLONIAN MATHEMATICS

We turn first to the Babylonian culture, or more precisely, that of Mesopotamia, where the Tigris and Euphrates rivers flow through a fertile valley. Here, where clay was found in abundance, primitive writing began. Known as cuneiform writing, it was done by impressing wedge-shaped marks with a stylus in soft, wet clay, which was then baked in ovens or dried in the heat of the sun. These clay tablets were permanent, and thousands of them, dating as far back as 4000 B.C., have come down to us today. Naturally, only some of these tablets involve mathematics. A tablet with cuneiform writing is shown on the left in figure 1.2.

Fig. 1.2. Cuneiform writing (*l.*) and number concept (*r.*)

We note next that the Babylonian system of numerals, although dependent on the repetition of symbols, was built on a base of 60 rather than a base of 10. The meaning of the symbols depended in part on their position when a number was written. Small vertical wedge-shaped marks indicated ones; somewhat larger horizontal wedge-shaped characters indicated tens. Thus

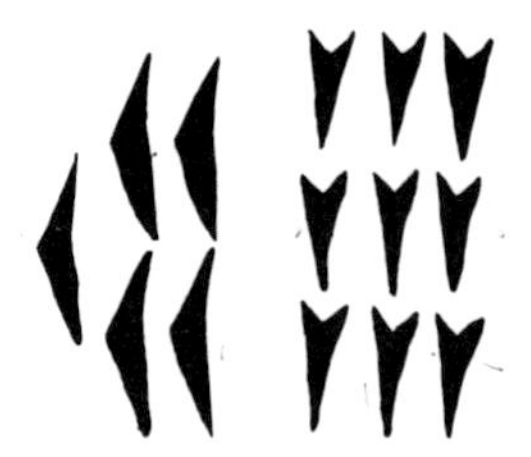

represented 59. Beyond 59, however, the system changed. The Babylonians somehow hit on the idea that the same symbol, ⟨, in different positions might readily represent 60, or $(60)^2$, just as well as 10. This so-called sexagesimal system was well suited to their astronomical calculations. Curiously, because of this "positional" feature, it is closer to our own numerical notation than most other ancient systems are.

The stamp of Nicaragua at the right in figure 1.2 might represent a Babylonian figure as well as an Egyptian figure. The stamp is the first of a set of ten stamps issued under the collective title Ten Mathematical Formulas That Changed the Face of the Earth. The message, "$1 + 1 = 2$," is intended to commemorate the evolution of the concept of number, calling attention to the basic ideas of counting and addition—ideas that can scarcely be attributed to any one particular culture. Ironically, the Arabic symbols used—"1," "2," and "+"—did not come into general use in Europe until several thousand years later, probably about the thirteenth century.

We also find tables of interest charges on loans. At first the Babylonians charged interest at the rate of from 20 to 30 percent; later, the interest rates varied from 5 to 25 percent for precious metals (i.e., coin) and from 20 to 33⅓ percent for produce. In addition to collecting interest on loans, the Babylonians were also well versed in the use of weights and measures.

The geometry of the Babylonians consisted essentially of a few empirical rules. For example, they knew that—

1. the area of a rectangle is the product of the length of two adjacent sides;
2. the area of a right triangle is equal to one-half the product of the lengths of the sides adjacent to the right angle;
3. the area of a trapezoid with one side perpendicular to the parallel sides is one-half the product of the length of this perpendicular and the sum of the lengths of the parallel sides.

The Babylonians were also acquainted with the "right-triangle rule," that is, the numerical aspect of the so-called Pythagorean theorem. Although they probably never "proved" this relationship in general terms, they did develop tables of Pythagorean triples, that is, triads of whole numbers that satisfy the relation $a^2 + b^2 = c^2$. For instance, they were aware that when $a = p^2 - q^2$, $b = 2pq$, and $c = p^2 + q^2$, then $a^2 + b^2 = c^2$, where $p > q$.

Other aspects of the Babylonians' knowledge of geometry probably included—

1. the area of a circle as $3r^2$ (a very crude approximation);
2. the perimeter of a circle as $6r$ (also very crude);
3. the volume of a prism and of a cylinder (the latter, approximate).

The Babylonians appear to have studied mathematics for its own sake as well as for practical purposes. This can be seen from their knowledge of elementary algebra, which, although somewhat limited, surpassed that of the Egyptians. For example, in addition to the Pythagorean triples, they were able to solve the following types of equations:

$$ax = b$$
$$x^2 = a$$
$$x^2 + ax = b$$
$$x^2 - ax = b$$
$$x^3 = a$$
$$x^2(x + 1) = a$$

They were also able to solve certain systems of equations in two variables, were familiar with some simple identities, and knew about the summation of arithmetical progressions.

But it was in astronomy that they excelled. Although the pseudoscience of astrology originated with the Babylonians, their knowledge of the heavens was substantial. They were acquainted with the planets Mercury, Venus, Mars, Jupiter, and Saturn, and they made extensive lists of stars. The major responsibility of the astronomers was to observe the moon carefully in order to keep the calendar. Basing their data on empirical observations, they were able to predict eclipses with remarkable accuracy. In this connection, it may be noted that contrary to general belief, the Babylonians divided the circle, not into 360 parts, but rather into 8, 12, 120, 240, and 480 parts. In all probability, we have to thank the Greek mathematician Hipparchus (ca. 50 B.C.) for dividing the circle into 360° in connection with the development of trigonometry. In any event, much of later Greek astronomy was built on the astronomical knowledge of the early Babylonians.

EGYPTIAN MATHEMATICS

The ancient Egyptians recorded events on four kinds of material: stone; papyrus, a crude form of paper fashioned from the pulp of a water reed; wood; and pottery. When writing on stone, they carved pictorial characters slowly and with great care; this type of writing is known as *hieroglyphic* writing. (The Austrian stamp on the left in fig. 1.3 portrays a typical Egyptian hieroglyphic stone carving.) For more rapid writing on papyrus, wood, or pottery, the Egyptians used free-flowing cursive scripts. One of these was the *hieratic* writing used by the priests; the other, *demotic* writing, was that used by ordinary people.

Fig. 1.3. Hieroglyphic stone carving (*l.*) and Rosetta stone (*r.*)

How do we know all this? When Napoleon sent an expedition to Egypt in 1799, his soldiers discovered the Rosetta stone in the Nile Delta. It was brought to England in 1801 (despite the protests of the French) and is now housed in the British Museum. This slab of basalt rock, about 70 centimeters by 112 centimeters and 27.5 centimeters thick, is the key that unlocked the puzzle of how to read the hundreds of ancient Egyptian papyri that still exist, enabling scholars to decipher hieroglyphics. The Rosetta stone contains inscriptions in three scripts: (1) Egyptian hieroglyphic picture writing; (2) Egyptian demotic, or popular, cursive writing; and (3) Greek. Although it was written only in 196 B.C., the stone (shown at the right in fig. 1.3) was a boon to scholars, who could now decipher Egyptian papyri many centuries older.

Our knowledge of Egyptian mathematics is obtained from two fairly well preserved papyri: the so-called Moscow papyrus, written about 1850 B.C., and the Rhind papyrus, written about 1650 B.C. The Rhind papyrus is about 5.4 meters long and 30 centimeters wide; the numerals are in cursive script. It is also kept in the British Museum.

The Egyptian stamp in figure 1.4, commemorating the International Congress of Medicine at Cairo in 1928, depicts the legendary figure of the Egyptian architect and royal physician Imhotep, who lived about 3000 B.C. Resembling the Greek god of healing, Aesculapius, Imhotep supposedly was familiar with the mathematics later incorporated in the Rhind papyrus.

Fig. 1.4. Imhotep

The Egyptian system of numeration is typical of some other systems of ancient cultures, notably the Roman, in that it uses the base 10 and depends chiefly on the repetition of a few symbols. Thus, in hieroglyphic numerals:

1 = a vertical stroke
10 = an inverted U, or horseshoe
100 = a coiled rope
1 000 = a lotus flower
10 000 = an upright bent figure
100 000 = a tadpole
1 000 000 = a god with uplifted arms

For example,

represented the number 245.

The arithmetic of the Egyptians was similar to that of the Babylonians except, of course, for their numerical notation. Some rudiments of algebra also seem to have appeared, but Egyptian algebra was limited to a consideration of simple linear equations, such as $x + 3x + 5x = 33$.

Although the Iranian stamp at the left in figure 1.5 commemorates a contemporary census count, it suggests that in ancient times finger reckoning was quite common. The stamp at the right depicts ancient scribes doing arithmetic, with a stylized abacus at the left of the design and stylized magnetic computer tapes at the right. (Such anachronisms frequently occur on commemorative stamps.)

Fig. 1.5. Finger reckoning (*l.*) and Egyptian scribes (*r.*)

The Egyptians were skillful in handling fractions, especially so-called unit fractions—that is, fractions with a numerator of 1, such as ½ or ¼. Of course they did not regard a fraction quite as we do today, nor did they use our familiar notation. Nevertheless, they discovered many properties of fractions. They would think of $\frac{3}{5}$ as the sum of $\frac{1}{3} + \frac{1}{5} + \frac{1}{15}$, or they would write $\frac{2}{13}$ as $\frac{1}{8} + \frac{1}{52} + \frac{1}{104}$. They compiled elaborate tables of such equivalences in manipulating fractions.

Interestingly enough, the Egyptians had a deep concern for astronomy. They observed that the Nile River overflowed its banks at very definite intervals of time in accordance with the positions of certain stars. From this, they devised as early as 4000 B.C. a solar calendar of 365 days, comprised of twelve months of 30 days each, with 5 extra feast days. The influence of the Nile is one of the most fascinating aspects of Egyptian mathematics. The river's annual flooding created problems for surveyors, who had to determine the boundaries obliterated by the inrushing and receding water, primarily for purposes of taxation. It is believed that the surveyors used ropes with knots placed at regular intervals to stake out a right angle by using the 3:4:5 ratio of the sides of a right triangle.

This brings us to an examination of Egyptian geometry, which was essentially empirical—that is, it was based on observation, intuition, and practical measurement. Their geometry was essentially applied arithmetic, with very little hint of any geometric generalizations or relationships. In connection with the mensuration of geometric figures, for example, the Egyptians considered that—

1. the area of an isosceles triangle was equal to one-half the product of the base and the altitude;
2. the area of an isosceles trapezoid was equal to one-half the sum of the bases multiplied by the altitude;
3. the area of a circle divided by its circumference was equal to the area of the circumscribed square divided by the perimeter of the circumscribed square;
4. the value of π was equal to $4(\frac{8}{9})^2$, which is equivalent to 3.1605, a fairly close approximation;
5. the volume of a frustum of a square pyramid and that of a circular cone could be found with formulas they had developed (both of which turned out to be incorrect).

In short, Egyptian geometry was a rule-of-thumb affair; the need for measurement was sufficient motivation, accompanied by a willingness to accept approximate results. Despite the crudity of the Egyptians' geometry, the Pyramids give further evidence of their skill in measuring as well as in the practical arts.

The three pyramids of Gizeh (fig. 1.6), near Cairo, are the largest and finest of their kind. Other cultures have imitated them, but the true Egyptian pyramids were royal burial tombs. They are masterpieces of engineering. The Great Pyramid of Cheops (Khufu), the largest ever built, is one of the Seven Wonders of the World. Built about 2900 B.C., it is an amazing piece of construction. It is said that a hundred thousand human beings worked on it for fifty years. More than two million stone blocks, weighing an average of over two tons apiece, were fitted together perfectly. Covering some thirteen acres, the base is very nearly a perfect square about 226

Fig. 1.6. The great pyramids

meters on a side; the pyramid towers 144 meters to its apex (the Washington Monument is 166 meters high). As if this were not impressive enough, within the pyramid are a number of chambers connected by passageways. The roofs of these chambers are huge granite blocks (8.1 x 1.8 x 1.2 meters) weighing over forty-five metric tons apiece. They were transported from a quarry 960 kilometers away, and then raised to a height of 60 meters above ground level by means of a brick ramp. We can only marvel at the ingenuity of these master builders. The Egyptians used their knowledge of proportion in order to construct the sloping sides of the Pyramids. They also knew how to construct special angles, as shown by the location of the chambers and passageways within the pyramidal tombs. Most of this knowledge and skill was jealously guarded by the priests and not shared with the common people.

Summing up the contributions of the Egyptians, we note that the valley of the Nile was a peaceful, sheltered area conducive to an introspective way of life. The golden age of Egyptian mathematics extended from 1900 to 1600 B.C. Thereafter, for whatever reason, followed a period of stagnation that lasted for the next thousand years or so.

Viewing both Egyptian and Babylonian mathematics in perspective, we clearly see that in neither culture was there any great concern about exact results as distinct from approximations. There was little or no desire to arrive at generalizations and little, if any, awareness about the nature of mathematical proof—nor, for that matter, about abstract concepts. As to the motives that gave rise to their mathematics, two alternative theories are possible. According to one viewpoint, geometry originated in Egypt from practical needs, such as measuring in connection with the Pyramids and surveying when restoring the land boundaries that were periodically obliterated by the flooding waters of the Nile. According to the other theory, it might be said that the priestly leisure class, with its rituals and ceremonies, encouraged the study of mathematics for its own sake and for its aesthetic aspects, as revealed by the symmetry and design in their pottery, weaving, and basketry. Probably the truth lies somewhere between the two, and the origins of mathematics were shaped by both considerations.

An interesting contrast can be drawn between the Babylonian and the Egyptian cultures. The Babylonians were more imaginative: they established cities and invented writing. Constantly on the alert against attacks from neighboring barbarians, they had little time for reflection. The Egyptians, however, were not as creative or as persistent as the Babylonians, and their contributions to civilization were fewer and less significant. Life in the valley of the Nile was too easy to challenge their inventiveness, and it is not surprising that the Egyptians became a reflective and contemplative people who turned to religion.

Thus Babylonia—and to some extent Egypt—laid the early foundations of civilization. From the Babylonians we have inherited the concepts of city life, defensive warfare, private ownership, and the habit of putting things in writing. From Egypt, other than superb masterpieces in the art of sculpture, we have inherited very little of lasting significance: no philosophy, no body of law—even their astronomy was destined to oblivion.

Still, the faint beginnings of what we today call *mathematics* could from time to time be discerned. Perhaps the stage was being set for the impressive achievements of the Greeks, hovering in the wings.

Philosophers and Geometers

600 B.C.–A.D. 400

THE "science" of the ancient Greeks would scarcely qualify as science by modern criteria. Their notions were derived chiefly from speculation and some little experimentation, not from objective observations and measurements (although they did recognize that the earth was a sphere). They toyed with ideas of the nature of matter and its indestructibility, but the atomistic theory of Democritus was quite different from that of Dalton some two thousand years later. Greek medicine did lean largely on experiments by Hippocrates, but Plato's physics and physiology were anthropomorphic. Aristotle, a keen observer, was more successful in physiology than in physics, but he is chiefly remembered for his logic.

Greek architecture was dignified and impressive, with a grandeur rarely surpassed, but like Greek geometry, it lacked dynamic quality: no flowing movement, only frozen symmetric beauty. This can be seen in the Ionic capital (the top, or crown, of a supporting pillar) at the left in figure 2.1. The other two types of capitals used by the Greeks were the Doric and the Corinthian. The Romans used somewhat similar capital designs. After the fall of Rome, these designs were modified by Gothic architects, but the classic forms were revived during the Renaissance and have survived to the present day. The modern Austrian parliament building at the right in figure 2.1, for instance, is typical of the architecture of ancient Greece.

In addition to their intellectual, philosophic, and aesthetic talents, the Greeks were fairly skillful in handicrafts and technics. They were familiar

Fig. 2.1. Ionic capitol and Greek architecture

with the potter's wheel and the lathe, and they knew how to solder iron. They used the *gnomon* (sundial) to tell time. They believed that the primary stuff of the physical world consisted of water, earth, air, and fire.

During the course of nearly a thousand years, mathematics flourished among the Greeks. We know that they contributed ideas to the theory of numbers, which they called *arithmetike,* but were much less interested in the art of computation, which they called *logistike*. The former was the concern of philosophers and scholars; the latter was relegated to merchants and artisans. The greatest single contribution of Greek mathematics to posterity was, of course, the concept of deductive reasoning, or logical proof, which was probably introduced by Thales (ca. 600 B.C.) and elaborated on by Euclid and his successors. Because of the association of logic with philosophy, the Greeks regarded geometry as an essential part of a liberal education. (In medieval universities the *quadrivium* consisted of arithmetic, geometry, astronomy, and music, whereas the *trivium* was composed of grammar, rhetoric, and dialectic; together they formed the seven liberal arts.)

We now know that some mathematical knowledge formerly believed to be original with the Greeks must instead be attributed to the Egyptians and to the Babylonians. Nevertheless, the glorious achievements of Greek geometry are not likely to fade even after two thousand years. As G. H. Hardy has so truly said, "The Greeks were the first mathematicians who are still 'real' to us today . . . Greek mathematics is the real thing."

THE RISE OF GREEK SCIENCE

The rise of Greek science began in Ionia, near the Aegean Sea, with Thales and Pythagoras. Of these two men we know very little directly, only what later writers said of them. It seems that Thales (ca. 600 B.C.) was a man of practical affairs, a shrewd businessman, and a philosopher of sorts, with an interest in astronomy. From his travels in Egypt he brought back some knowledge of Egyptian geometry. He is credited with a few elementary theorems, but it is doubtful if he did more than take the first timid steps toward organizing geometry on a systematic basis.

Even more mystery and legend surround his compatriot Pythagoras (ca. 540 B.C.), for nearly all we know of him has come down through second- and third-hand sources. Without doubt Pythagoras was a mystic and a prophet. In his rather wide travels he gathered considerable knowledge of mathematics and astronomy as well as religious lore. On settling in Croton, a city in southern Italy, he organized a secret society of kindred spirits interested in philosophy and mathematics. In time this Pythagorean brotherhood became firmly established, with elaborate rules and rituals. Pythagoras not only taught the members of this "inner circle" but lectured to the general public as well. Credit for each new discovery was generally given to Pythagoras himself, not to individual members of the brotherhood.

Much of what was once regarded as contributions of Pythagoras, including the so-called Pythagorean theorem itself, had actually been known earlier by the Babylonians. It is even possible that the Chinese may have been familiar with the Pythagorean relation (see *Was Pythagoras Chinese?* by Frank J. Swetz and T. I. Kao [University Park, Pa.: Pennsylvania State University and NCTM, 1977]).

Yet the Pythagorean school gave mathematics in Greece an entirely new complexion. Mathematics was henceforth regarded as closely related to philosophy and wisdom in general. And arithmetic (number theory, not computation) was regarded by the Pythagoreans as even more significant than geometry. Pythagoras is supposed to have said, "Number rules all." In fact, the Pythagoreans were among the first to use geometry to express relations between quantities and numbers. They developed a theory of proportion that served them well until they encountered numbers that could not be expressed as the ratio of two whole numbers, such as $\sqrt{2}$ and $\sqrt{40}$.

In any discussion of the Pythagoreans, it is difficult to separate legend from fact. However, the most notable aspect of the Pythagorean influence is the belief it fostered in the universal importance of numbers. Clinging to number mysticism and worship, they assigned all sorts of characteristics to numbers. Odd numbers had male attributes; even numbers, female traits. The number two was the first female number; three was the first male number. Four was the number of justice; five represented marriage; and so on. This sort of idolatry led to numerology and astrology, which still enjoy popularity today. Of far greater significance mathematically was the

Pythagorean discussion of figurate numbers, which leads not only to interesting geometric patterns but also to infinite progressions of numbers (see fig. 2.2).

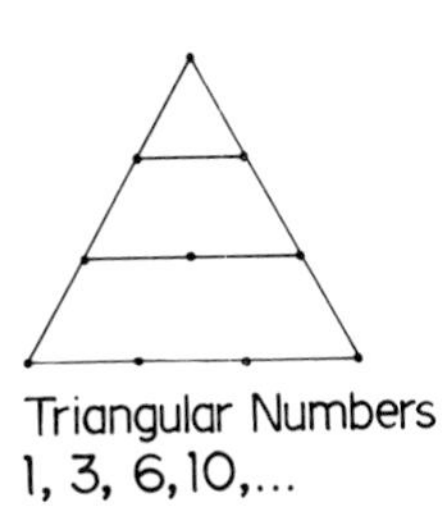

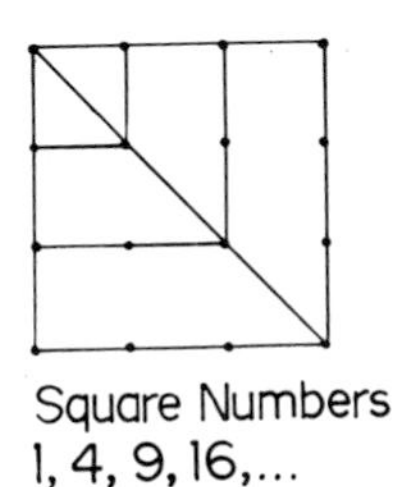

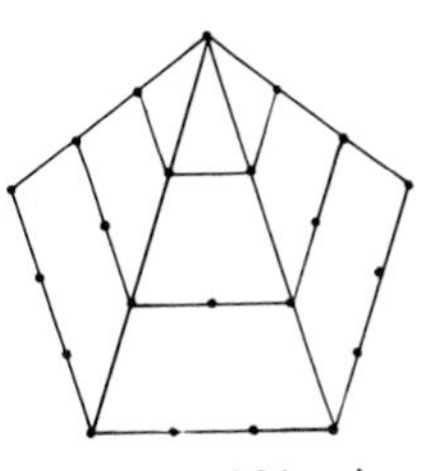

Pentagonal Numbers
1, 5, 12, 22,. . .

Fig. 2.2

Of interest also is the five-pointed star, or pentagram, which was the badge of the brotherhood and which led to a greater knowledge of geometry. This included the golden section, which refers to the division of a line segment into two parts such that the ratio of the length of the entire segment is to the length of the greater part as the greater part is to the smaller part (fig. 2.3). Either intersection point (e.g., *B* or *C*) on any diagonal of a regular pentagon divides that diagonal into the golden section.

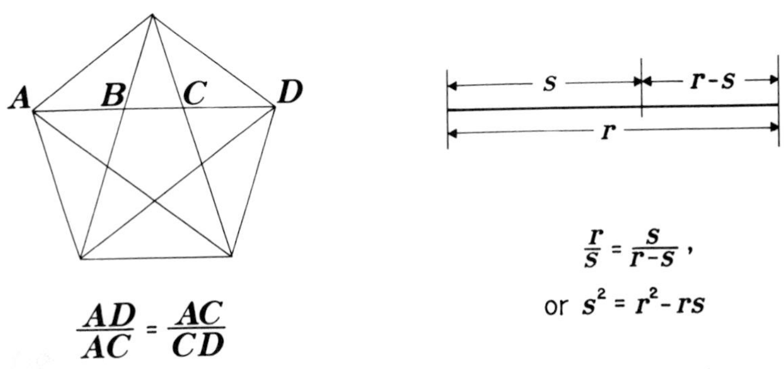

Fig. 2.3

We come now to the famous right-triangle theorem itself. Both Pythagoras and his theorem have been honored on stamps. At the left in figure 2.4 is an artistic representation of the theorem for the special case of a 3:4:5 right triangle; the stamp at the right depicts a Greek coin from the island of Samos, birthplace of Pythagoras, showing Pythagoras presumably consulting an oracle; in the center another stamp shows a geometric

Fig. 2.4. Pythagorean theorem

diagram of the Pythagorean relation. A statement of the theorem, $A^2 + B^2 = C^2$, is given on the stamp in figure 2.5, which nicely illustrates Greek architecture once more.

Fig. 2.5. Law of Pythagoras

It is not clear how the Pythagoreans justified the theorem, which, as we have already said, was known to the early Babylonians and Egyptians. What "geometric proof," if any, Pythagoras gave is not known, but Euclid and others gave several proofs. Since the days of Greek geometry, over a hundred mathematical proofs of the theorem have been given, some of them geometric, others largely algebraic. Two are suggested in figure 2.6; readers might like to see if they can figure them out by themselves.

Next, we turn to the stage of Greek mathematics known as the "Athenian school," which ranged over the fifth and fourth centuries B.C. Thales and Pythagoras had opened the way, laying the foundations of geometry and theoretical arithmetic. Now the city of Athens was thriving

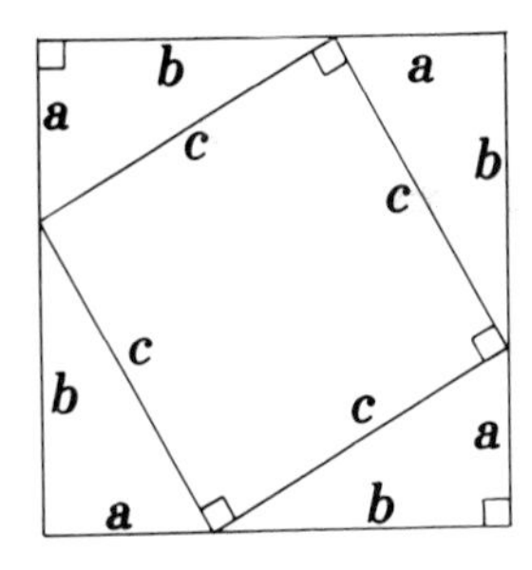

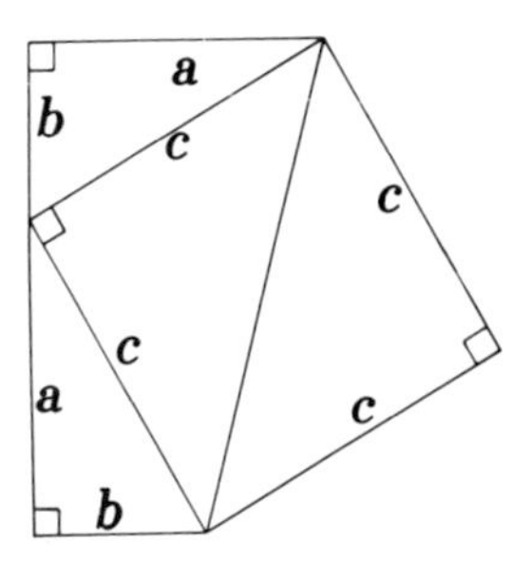

Fig. 2.6.

as the intellectual center of the Greek world. Philosophers, mathematicians, and astronomers gathered in Athens, among them Democritus, Plato, and Aristotle.

Democritus struggled with the problem of dividing a solid such as a cone or a cylinder into many thin slices, an idea that anticipated the development of the calculus many centuries later. His thinking suggested the doctrine of atomism. According to this notion, all physical phenomena could be explained in terms of extremely small but finite, hard atoms, incapable of subdivision and moving about in empty space. He believed these atoms were unchangeable and indestructible. The atoms differed from one another in size and form and properties, which accounted for the variation in the properties of different substances.

Democritus is honored on the stamp at the right in figure 2.7, which also bears the contemporary symbol of atomic structure; it was issued to commemorate the inauguration of the Democritus Nuclear Research Center at Aghia Paraskevi, in Greece. Democritus's concept was remarkably similar to modern theories of the structure of matter, but whereas contemporary physics and chemistry are based on quantitative experiments and

Fig. 2.7. Aristotle and Democritus

rigorous mathematical analysis, the doctrine of Democritus grew out of intuition and philosophic speculation. Although his atomism was decried by Socrates and others, his views never died out completely. They were revived, so to speak, about 1800 by John Dalton, the English chemist, who built his theory on a century and a half of chemical experimentation.

A generation or so after Democritus, two other philosophers helped to mold Greek thought—Plato (ca. 380 B.C.) and Aristotle (ca. 340 B.C.). Plato was more deeply interested in *moral* philosophy than in *natural* philosophy (i.e., physical science), which he regarded as somehow inferior. He viewed mathematics as a more or less lofty, abstract form of thought, remote from the mundane affairs of everyday life. The single exception that Plato conceded was the relation of mathematics to astronomy. In his view, the heavens reflected the perfection of abstract mathematics, and this conviction prevailed for centuries until the time of Johannes Kepler (1570–1630).

Aristotle is honored on the left-hand stamp in figure 2.7. A great philosopher and an able biologist, he is also known for his systematic treatment of deductive logic—that is, reasoning from accepted statements to necessary conclusions. This elaborate Aristotelian *verbal* logic exerted profound influence until the nineteenth century, when the modern mathematician George Boole introduced the study of *symbolic* logic to mathematics. Aristotle tended toward experimental science; he rejected the atomistic speculations of Democritus. Aristotle's philosophy in his own day was not as influential as that of Plato, but the Arabs later "rediscovered" him, and for centuries thereafter Europe regarded the authority of Aristotle as the last word.

THE GOLDEN AGE

The period from 300 to 200 B.C. is often called the golden age of Greek mathematics. During this time creativity reached its zenith. Three men towered high above their contemporaries as well as above those who had preceded them or who were to follow: Euclid, Archimedes, and Apollonius.

Euclid

With the founding of the city of Alexandria about 330 B.C., mathematical activity shifted from Greece to Egypt. Although Alexander the Great did not live to see it, the metropolis that he founded at the mouth of the Nile blossomed into a renowned center of learning to which both Greek and Oriental scholars flocked in great numbers. In the course of time, the university at Alexandria boasted more than 700 000 volumes. One of the first teachers at the university was Euclid. His influence was felt a generation later in the brilliant achievements of two other scholars at the university, namely, Archimedes and Apollonius, probably the greatest mathematicians of antiquity.

Euclid's greatest contribution was his work entitled the *Elements,* in reality a series of thirteen books. In this he collected and systematized all the mathematics known at the time (about 300 B.C.). Books 1, 2, and 4 deal with lines and plane figures (due chiefly to Pythagoras); Book 3 deals with circles (Hippocrates); Book 5 is devoted to the theory of proportion (Eudoxus); Book 6, areas and similar figures; Books 7, 8, and 9, the theory of numbers (mostly Pythagoras); Book 10, irrational quantities; Book 11, solid geometry; Book 12, the method of exhaustion; and Book 13, proofs and constructions for the five regular solids (Plato).

Despite certain logical weaknesses, this monumental work survived virtually unchanged for more than two thousand years. Indeed, the *Elements* was so revered that in England many centuries later, instead of studying geometry, one simply "read Euclid." The logical weaknesses alluded to chiefly concern the postulates, particularly the so-called parallel postulate. These shortcomings were eventually remedied in the middle of the nineteenth century, but that story must wait until a later chapter.

Archimedes

Although Archimedes (ca. 250 B.C.) was without doubt the greatest scientist and mathematician of ancient times, he also achieved fame for his "practical" discoveries and ingenious mechanical devices. He was the first to explain the law of the lever and was so thrilled by its usefulness that he is supposed to have exclaimed, "Give me a place to stand on and I can move the world." Actually, in recognizing that the weights and lever arms are in inverse proportion, he applied quantitative measurement to observations, anticipating the science of mechanics by nearly two thousand years.

Archimedes is honored on the two stamps shown in figure 2.8: at the right is a reproduction of a painting by the seventeenth-century artist José

Fig. 2.8. Archimedes and his law

Ribera as he imagined Archimedes; at the left is a beam balance illustrating Archimedes' law of the lever, which is given here as $F_1x_1 = F_2x_2$.

The story of the gold crown, which led to the discovery of the principle of buoyancy, is well known. Archimedes had been asked to determine, without damaging the crown, whether it was made of pure gold or not. On entering his bath one day, he noted that when an object is immersed in a liquid, the volume of liquid displaced equals the volume of the object itself. He could compare the volume of the crown with the volume of an equal weight of pure gold; if the volumes were equal, the crown would be pure gold. If, however, the "gold" crown were mixed with silver, it would have a greater volume, since silver is bulkier than gold. In the excitement over his discovery, he dashed naked through the street crying, "Eureka! Eureka!"—"I've found it! I've found it!" Indeed, he later proved by this method that the crown was not of pure gold.

Among the many mechanical devices attributed to Archimedes were the endless screw, which served as a water pump; huge lenses, or "burning glasses," with which he could set fire to ships at sea; the cranes and systems of pulleys that enabled him to lift a ship out of the water; tremendous catapults and war engines, which he used against the invading Roman armies at the siege of Syracuse.

Presumably Archimedes thought less of his mechanical ingenuity than of his purely mathematical achievements. These concerned the calculation of the areas, volumes, and centers of gravity of various curves, surfaces, circles, spheres, conics, and spirals. He also calculated a value for pi (π), that is, the ratio of the circumference of a circle to its diameter. He was able to prove that the value of π lies between 223/71 and 220/70. This works out to be between 3.1408 . . . and 3.1428 . . . , or a mean value of 3.1418, which is very close to the commonly used modern approximation 3.1416 and was closer than any other value known to the ancient world (with the possible exception of the Chinese). Today, because of the miraculous power of electronic computers, we know the value of π correct to more than 100 000 decimal places. This figure is completely useless for most practical purposes of computation (at least beyond the first dozen or two digits), but it is of considerable interest in the theory of numbers. The method used by Archimedes to arrive at his results was to calculate the perimeters and diameters of regular polygons drawn inside and outside a circle. As the number of the sides of the polygons was increased, the perimeters of the inner polygons grew longer and those of the outer polygons grew shorter. The circumference of the circle, remaining constant, was trapped between them. By using polygons of ninety-six sides, Archimedes arrived at the two limiting values given above.

Another significant contribution by Archimedes was his famous sand reckoner, with which he attempted to calculate the number of grains of sand required to fill the entire universe. To achieve his purpose, he devised a scheme for expressing very large numbers that is virtually the equivalent of modern exponential notation. The significance of this pioneer attempt is that it showed that *anything that really exists, no matter how large, can be counted or measured.*

Apollonius

Apollonius was regarded in his day as "The Great Geometer," chiefly on account of his remarkable studies of the conic sections—the circle, the ellipse, the parabola, and the hyperbola. These results paved the way for later use by Kepler in his astronomy, just as Archimedes' calculations of areas and volumes anticipated the calculus of Newton (ca. 1790). In retrospect, it would seem that if either of these ancient geometers had had available an effective algebraic symbolism, they might conceivably have achieved what was not to come until some two thousand years later. The history of science abounds with such instances of discoveries or inventions that "almost made it" centuries before their time but failed only because some necessary tools were lacking.

Regrettably, neither Euclid nor Apollonius has ever been honored philatelically.

THE DECLINE OF GREEK MATHEMATICS

Hipparchus

In the closing centuries before the beginning of the Christian era, we find the Greek astronomer Hipparchus (ca. 150 B.C.) insisting that the planets revolved around the earth in a complicated system of circles, with the unmoving earth as the center of the universe. His system worked when it came to calculating the positions of the planets, and so, in keeping with the philosophy of Plato and Aristotle, astronomers clung to this theory for over fifteen centuries, until the time of Copernicus (1543).

Hipparchus is shown on the smaller stamp in figure 2.9, together with a simplified armillary astrolabe to commemorate his work in astronomy. Among the other notable contributions by Hipparchus must be counted his work in trigonometry, especially his careful computations and compilation of a table of values of corresponding arcs and chords of a circle for an entire series of angles. It is probably owing to his influence that the circle was divided into 360 degrees. His brilliant work also included the drawing of the first accurate star map, which recorded the exact positions of more than a thousand of the brighter stars—no mean achievement without a telescope!

Ptolemy

During the first four centuries of the Christian era, Greek mathematics gradually deteriorated. It no longer flourished as it had in the earlier days of the Alexandrian school or the brilliant era of Archimedes and Apollonius. To be sure, Claudius Ptolemy (second century A.D.) substantially agreed with Hipparchus's conception of the planetary system; the earth was still regarded as the pivot around which the planets moved, some now supposedly in eccentric circles.

Ptolemy not only embraced the ideas of Hipparchus but also added his own ideas, and in time they came to be known as the Ptolemaic system. After the fall of the Roman Empire, his writings survived among the Arabs, who translated them in a book called the *Almagest*—later translated into Latin (1175). The Ptolemaic system dominated European astronomical thinking through the Renaissance until the time of Copernicus.

Ptolemy has been commemorated on a huge square stamp of the Yemen Arab Republic, which also depicts, by way of contrast, the contemporary American telecommunication satellitle *Vanguard 1* (fig. 2.9).

Fig. 2.9. Hipparchus and Ptolemy

The Romans

In closing our remarks on mathematics in classical antiquity, we observe that the Romans were less skillful in computing than the Greeks, although they used computations constantly in surveying and engineering and in the construction of their roads and aqueducts. To aid them in their computing, they devised three forms of the abacus: (1) a primitive dustboard, (2) a grooved table with sliding beads, and (3) a table marked in columns to be used with counters.

The familiar numerals used by the Romans appear on literally hundreds of stamps, notably stamps concerned with religious subjects and stamps depicting Olympic games. The stamp at the left in figure 2.10 has the date 1818 written as MDCCCXVIII, which is the way the Romans probably would have written it, to commemorate the 150th anniversary of the inauguration of the Prague National Museum. On the Vatican City stamp at the right we find the year 1963 expressed in Roman numerals the way it is generally done in modern times, namely, MCMLXIII rather than MDCCCCLXIII. This stamp also gives the day and the month in Roman numerals, commemorating the coronation of Pope Paul VI on 30 June 1963.

Fig. 2.10. Roman numerals

Presumably the Romans did not ordinarily use the subtractive principle. They did not write IV for 4 or IX for 9; instead, they wrote IIII for 4 (which is still seen on some clocks and watches) and VIIII for 9. The use of Roman numerals in Europe persisted long after scholars stopped writing in Latin. Even today we still find them on tombstones, monuments, cornerstones of buildings, and on books, magazines, and films.

Except for improved forms of the abacus and for their numeration system, the Romans added precious little that was new to mathematics. Indeed, the Romans transmitted only a little of Greek mathematics to the Middle Ages. It was not until six centuries later that information about Greek science and mathematics was made available to European scholars through the translation of Arab works into Latin.

The Greeks had given the world a priceless heritage: they had contributed two concepts that were to influence mathematics profoundly until the beginning of the present century. One of these was the unshakable faith in the method of deductive reasoning. The other was the firm conviction that our physical environment could be adequately described in terms of mathematics. To all intents and purposes, these two ideas are still vital in contemporary thought.

While Greek mathematics held sway, momentous events were taking place elsewhere. The Chinese made paper for the first time about 200 B.C., about the same time that they began to build the Great Wall; Caesar conquered Gaul about 50 B.C.; the great library of Alexandria was burned in A.D. 400; and the Christian era was ushered in. Rome fell in A.D. 476, and although Alexandria survived until the Arabs took over about 640, Greek knowledge of mathematics and science was all but lost to western Europe. The darkness of night had set in. It was to last for a thousand years.

Physicians and Clerics

A.D. 400–1500

At THE very time that the Roman Empire fell, a few scholars in Italy, chiefly Boethius, were writing texts based on Greek mathematics. During the Middle Ages, others added their meager contributions to the field: Alcuin, Gerbert, and Thomas Aquinas. But mathematically their activities were of no great significance—more important were the teachings of Adelard of Bath (ca. 1120) and Robert of Chester (ca. 1140), who, during their travels in the eleventh and twelfth centuries, served to hasten the awakening of mathematics in Europe. Adelard was one of the first to translate Euclid into Latin, and Robert of Chester translated al-Khwarizmi's algebra into Latin.

Mathematically, European minds were merely dormant, not sterile.

A PERIOD OF TRANSITION

Meanwhile, the Mohammedans successively conquered Jerusalem, Egypt, Persia, and finally Spain. The Moslems and Hindus had a flair for algebra and computation, but they showed little interest in geometry and logic. Although some Hindu arithmetic and algebra was original, much of it was borrowed from the Greeks. Hindu mathematics was translated by the Moslems into Arabic and Persian, and thus it finally found its way into Europe. While Europe stagnated, Moslem scholars translated all the

Greek classics they could find—on mathematics, science, philosophy, and medicine. The overall influence of Near Eastern scholars on mathematics was that of preservation and transmission. Nothing terribly important was created. The elementary algebra and the plane and spherical trigonometry that were handed down are of some interest to contemporary mathematicians, for they served chiefly as catalysts to reawaken and redirect mathematical thought in Europe around the beginning of the fourteenth century.

With the rise of the great universities at Salerno, Paris, Oxford, Cambridge, Padua, Naples, Bologna, and elsewhere about 1200, Europe entered a period of transition and gradual awakening.

The stamp in figure 3.1 was issued to commemorate the 500th anniversary of the founding of the University of Freiburg, in Baden-Württemberg, West Germany, a cultural center since medieval days.

Fig. 3.1. Medieval university

The beginning of intellectual freedom and the revival of learning brought about by the universities ushered in the Renaissance. Literary and artistic activities in Italy were triggered by a rediscovery of classic Greek ideals and philosophy as well as the impact of new ideas filtering in from the Byzantine Empire.

We see Thomas Aquinas (ca. 1250) patiently trying to reconcile theology and philosophy and Roger Bacon (ca. 1250) courageously defending experimentation and mathematics as essential for scientific advance.

Much of Renaissance architecture was based on mathematical ratios suggested by Pythagoras and Plato. Following the Roman architectural theorist Vitruvius, fifteenth-century architects seized on the notion that the proportions of buildings should be derived from those of the human body. Painting and sculpture were represented by the work of Botticelli, Raphael, Michelangelo, Albrecht Dürer, and the immortal Leonardo da Vinci.

The thirteenth and fourteenth centuries saw the advent of mechanical clocks and the use of wool and silk textiles from Spain. The greatest single contribution was Johann Gutenberg's invention of printing from movable type about 1450. The advent of the printing press has been commemorated on a number of stamps. A portrait of Gutenberg can be seen on the stamp at the left of figure 3.2; at the right is a seventeenth-century hand printing press. Throughout this entire period there was great interest in geography and navigation, culminating in the discovery of the New World by Columbus in 1492.

Fig. 3.2. Gutenberg and early printing press

Through the influence of Fibonacci's *Liber abaci* (1202), Hindu-Arabic numerals gradually began to replace Roman numerals. From 1500 onward, the numerals used were virtually the same as those we use today. Practical arithmetic and computational methods were further developed by Luca Pacioli, Robert Recorde, and Adam Riese in the early sixteenth century.

Toward the end of the Middle Ages, algebra and trigonometry were gradually cultivated in Europe. The close of the Renaissance saw the introduction of symbolic algebra by François Viète (or Vieta). Two of Viète's notable contemporaries were Thomas Harriot, who promoted effective algebraic notation, and Simon Stevin, who developed the theory of decimal fractions.

MATHEMATICS IN THE NEAR EAST

Arabic mathematics flourished during the ninth century. One of the earliest noted Arabic scholars was al-Kindi, who died in A.D. 870. He was known as the Philosopher of the Arabs, and his full name was abu-Yusuf Ya'qub ibn-Ishaq al-Kindi; in medieval Europe he was called Alchindi. He wrote more than two hundred treatises dealing with many subjects, including astronomy, astrology, medicine, optics, and number theory. He is shown in figure 3.3 at the left. Another noted Arabic scholar was al-Farabi *(right),* who flourished in Damascus during the first half of the tenth century. Highly thought of as a philosopher by his contemporaries, he wrote an elaborate treatise based largely on the work of Aristotle. He was one of the first Moslem scholars who tried to reconcile Aristotle's approach and the Islamic viewpoint. In mathematics he wrote a commentary on Euclid.

Fig. 3.3. al-Kindi and al-Farabi

Of greater significance was the work of the Arabian mathematician al-Khwarizmi, who lived from about 780 to 850. Among his contributions were two that have come down to the present. One is that he took ideas from both Greek and Hindu sources, including the Hindu numerals, together with the zero. When his works were translated into Latin, those numerals eventually became commonplace in Europe and to this day are called, somewhat inaccurately, Hindu-Arabic numerals. The other contribution was his famous book *Ilm al-jabr w'al muqabalah,* based on the work of Diophantus. When translated, the title means "the art of transposition and cancellation"; it applied to the procedures used in solving equations. The Arabic word *al-jabr,* meaning transposition, eventually became our word *algebra.*

Another important scholar of this period was al-Haytham (965–1039), known in Europe as Alhazen, an able mathematician and student of physics. He elaborated on some of the findings of Archimedes that had to do with solids of revolution. His most brilliant contribution was a book called *Treasury of Optics,* which was inspired by some of Ptolemy's discussions on the reflection and refraction of light. In fact, he delved deeply into the subject and carefully considered the structure of the eye. Often called the Father of Modern Optics, al-Haytham (commemorated on the two stamps in fig. 3.4) was one of the first to realize that we see an object because each point of it receives a ray of light and then reflects the ray into the eye. The ancients could never understand why an object seems to change in size when it moves toward or away from the observer, but Alhazen explained it as follows: The nearer an object is to an observer, the larger

Fig. 3.4. al-Haytham

will be the base of the cone of light rays that leaves the object and converges to the eye; hence the closer the object is to the observer, the larger it appears. This concept of a cone of light rays from an object to the eye is the principle on which the theory of perspective is based. Scientists paid little attention to Alhazen's idea for six hundred years, but artists like Albrecht Dürer and Leonardo da Vinci anticipated the scientists.

Many Arab scholars were not primarily mathematicians. So it was with Avicenna, the leading scholar and scientist in Islam, who was interested chiefly in medicine and philosophy. Recognized as the most learned sage of his day (980–1037), Avicenna, whose full name was abu-Ali al-Husayn ibn-Sina, was a child prodigy who reputedly could recite the entire Koran at the age of ten. Two stamps, among several others, have been issued to commemorate Avicenna (fig. 3.5). Avicenna's "book of recovery," which dealt with physics, logic, and metaphysics, helped to preserve the views of Aristotle among European scholars for many years thereafter. His *Canon,* translated into Latin, became the standard textbook of medicine until the seventeenth century. Avicenna, a prolific writer, produced more than a

Fig. 3.5. Avicenna

hundred books, the most significant of which were on medicine. The medical theory they expounded was based on the concepts attributed to Hippocrates and Galen. Interestingly enough, Avicenna was one of the few alchemists of his time who believed that the transmutation of base metals into gold was an impossibility.

Not to be overlooked in connection with the influence of Oriental thinkers is the Persian poet Omar Khayyám (ca. 1050–1122), whom his compatriots regarded more as a scientist because of his interest in astronomy and mathematics. He wrote a book on algebra in which he explored quadratic and cubic equations more deeply than al-Khwarizmi had done, offering both geometric and algebraic solutions.

The last of the Eastern writers we shall consider is the Hebrew scholar Moses ben Maimon (ca. 1135–1204), better known as Maimonides. Born in Spain, he spent most of his life in Egypt, where he became the personal physician to the renowned ruler Saladin. His chief claim to fame was his attempt, through his writings, to reconcile Aristotle's philosophy with the teachings of the Old Testament. He was also concerned with astronomy and the calendar. Maimonides is pictured on both stamps in figure 3.6.

Fig. 3.6. Maimonides

EUROPE AT THE CROSSROADS

We now turn our attention back to western Europe, still in an era of intellectual darkness. Nevertheless, a few scholars were devoting part of their time to mathematics. One of the most colorful of these was Gerbert (ca. 1000), a highly respected scholar in his time. We must remember that the Greek classics had found their way into Europe through a number of different channels: through the Arabs in northern Africa, through Sicily, and through the Moorish universities in Spain.

Gerbert, one of the first scholars to study in Spain, brought the Hindu-Arabic numerals into Europe. Born in France, educated in Spain and Italy, Gerbert was active in Germany as advisor to the Holy Roman Emperor Otto III. After serving as archbishop, Gerbert was elevated to the papacy and became Pope Sylvester II. He has been commemorated on the two stamps shown in figure 3.7, the one on the right showing him as pope with Archbishop Astrik.

Fig. 3.7. Gerbert, Pope Sylvester II

Gerbert wrote on arithmetic, astrology, and geometry. In the first two instances he borrowed from Arabic sources; his geometry was based largely on methods used by Roman surveyors. He also built clocks and astronomical instruments, again based on knowledge gleaned from Arab sources in translation. Actually, in this period a scholar would really have to have a good knowledge of Arabic.

Another ecclesiastic who contributed to the introduction of mathematics and science into Europe was Nicholas of Cusa (1401–1464), the son of a fisherman, born at Cusa (Kues) near Trier, Germany. He was regarded as a brilliant student by his contemporaries at the University of Padua. He was influential in bringing many Greek scholars to Italy and became an outstanding philosopher and theologian. At the age of twenty-five he was made a papal legate; later he became a bishop and eventually a cardinal. Often called the Divine Cusanus, he was a better cleric than mathematician, being guided more often by intuition than by logical thinking. As deputy to Pope Pius II, Cusanus had the leisure and peace of mind to devote his thoughts to the relation between cosmology and mathematics. For example, he held that the universe was infinite and could therefore have no center. He anticipated Copernicus in his belief that the earth not only rotated on its axis daily but also revolved about the sun. The trouble was that in the main his colleagues did not agree with his astronomy, and so his influence in that field was of little significance. His best-known treatise, "On Learned Ignorance," was a plea for simple truths as opposed to the verbal complexities of Aristotelian dogma.

Nicolaus Cusanus was also a jurist, astronomer, geographer, physician, scientist, and author. Interested in reforming the calendar, he proposed changes that would have made it similar to today's Gregorian calendar, which was not adopted until about a century after his death. Cusanus is also supposed to have prepared the first usable map of central Europe. In mathematics, he concerned himself with finding a value of π, arriving at 3.142337, quite close to the mean value 3.1418 of Archimedes. But he came to grief with the problem of constructing a square equal in area to a given circle. This he tried to do by averaging in some peculiar way the areas of the inner and outer polygons of a circle. Although he did contribute some ideas to the theory of numbers, his notions of measurement were misguided, and his attempts to clarify the mathematical concept of infinity fared little better. Cusanus is shown at the right in figure 3.8.

Fig. 3.8. Adam Riese and Cusanus

The beginning of the sixteenth century witnessed a surge of interest in mathematics in Germany. Many popular books on algebra and arithmetic appeared, among them several *Rechenbücher* ("arithmetic books") by Adam Riese (ca. 1520), the celebrated *Rechenmeister,* who is portrayed on the left in figure 3.8.

In particular, his *Rechnung auf der Linien und Federn* ("Reckoning on the Lines and Pen") explains methods of computation with counters on a board ruled with lines, as well as computation with the pen, that is, using the new Hindu-Arabic numerals. This was significant for two reasons: (1) it reflected the transition from counter reckoning, a manipulative procedure based on the principle of the abacus, to the more sophisticated use of numerals; and (2) it served to popularize and spread the use of the new Hindu-Arabic numerals, which had only recently been introduced into Europe. Using numerals to compute by arranging the work in a definite way is called an *algorism.* During this transition period, contests were frequently held between *abacists* and *algorists,* presumably to decide on the merits of the "new" method.

The latter half of the fifteenth century marks the beginnings of the intellectual reawakening called the Renaissance. A new world was unfolding. Scholars were throwing off the shackles of the Middle Ages; they were repudiating the speculations the ancients had handed down about human nature and the environment; and they were beginning to rely on their own observations and judgments. Science was beginning to shake off the influence of the alchemists and the mystics.

Upon this scene entered two men, Leonardo da Vinci and Albrecht Dürer. Neither was a mathematician; both were artists, but with a mathematical outlook. We usually think of Leonardo as one of the world's great painters, but he was far more than an artist and sculptor, achieving fame as an engineer, architect, inventor, scientist, and poet. He was a creative genius whose imagination and technical insight seem inexhaustible, and many of his ideas and inventions were ahead of their time. Although not professing to be a mathematician, he nevertheless frankly acknowledged the need for mathematical ideas in connection with his drawings and paintings. His feeling toward geometry was revealed not only in his art but also in his interest in mechanics, machines, optics, mathematical curves, star polygons, and, in particular, his emphasis on the principles of perspective. The list of his inventions is truly amazing: the centrifugal pump, a breech-loading cannon, rifled firearms, the universal joint, the conical screw, a rope-and-belt drive, link chains, roller bearings, a submarine, bevel gears, spiral gears, military tanks, the parachute, and even a flying machine, to mention but a few. Some of these devices were actually built, or at least models were made; others never went beyond the stage of a sketch. Ironically, their influence was not widely felt in his time, for he kept his ideas and sketches in notebooks. The comments accompanying the sketches were in mirror writing, which is not easily read without the aid of a mirror. Because Leonardo was left-handed, mirror writing came easily and naturally to him. He did not intend it as a secret handwriting; he was simply addressing an imaginary reader and did not feel it necessary to use conventional handwriting. The contents of his notebooks did not become generally known until many years later. Only recently a sketch of a chain-operated bicycle closely resembling a modern two-wheeler was found among his notes.

Leonardo da Vinci has been honored on scores of stamps. Two in particular—the Hungarian stamp at the upper left and the French stamp at the lower right in figure 3.9—show him as he is usually portrayed.

The German painter and engraver Albrecht Dürer, a contemporary of

Fig. 3.9. Leonardo da Vinci and Albrecht Dürer

Leonardo's, also revealed a blending of artistic talent and mathematical ability. Dürer, too, has been recognized on stamps; so have many of his paintings. Two such paintings, both self-portraits, are shown in figure 3.9—one as a young man *(upper right),* and the other in his more mature years *(lower left).*

Dürer's interest in mathematics was chiefly geometrical. He concerned himself with higher-plane curves, such as the epicycloid; he explored the relation of polyhedrons to their nets; and he gave several geometric constructions for the regular pentagon. Like Leonardo da Vinci, he devoted himself assiduously to problems of perspective and vision. In connection with geometric transformations and projections, Dürer prepared the way for Mercator, although this work was scarcely appreciated by his mathematical contemporaries. In addition to a book on applied geometry and a treatise on fortifications, Dürer also wrote on a rational system of perspective, on anatomy, and on the proportions of the human body. One of his well-known paintings, entitled *Melancolia,* portraying the spirit of the age, includes a familiar magic square, as shown in figure 3.10. The date of the painting, 1514, appears in the bottom row of the magic square.

16	3	2	13
5	10	11	8
9	6	7	12
4	15	14	1

Fig. 3.10. Magic square

We conclude our remarks on these two great painters by suggesting the significance of their work on the theory of perspective. Before the Renaissance, painters had freely disregarded visual perception. In other words, medieval art lacked the third dimension; the pictures were "flat," and the relative sizes and positions of objects were frequently unrealistic. The new trend, exemplified by Leonardo da Vinci and Albrecht Dürer, succeeded in depicting the subject of a painting as the viewer would see that subject in real life. This effect was achieved largely, though not entirely, by the use of a mathematical principle known as *focused perspective*. Perhaps you recall the drawing lessons of your school days: the horizontal reference line, the principal vanishing point, the vanishing points to left and right, actual parallel lines in the scene drawn so that they converge. By such means, depth and realism were achieved in drawings and paintings. The position of the artist's eye when painting a scene is all-important. Early Renaissance painters thought of their canvas as a glass screen interposed between themselves and the scene they were painting. Lines from the eye to points in the scene intersected this screen, forming an image of the scene. This image or section is known as a *projection*.

In short, the "secret" of perspective is that what we see is not static or frozen but rather is in a fluid, ever-changing state. Painters used a grid and a sighting device that enabled them to follow (in the mind's eye) the changing shape of an object, such as the right angles that become acute angles and the circles that become ellipses.

Looking back, we see that the period A.D. 400 to 1300, known as the Middle Ages, was indeed a dismal era of history. There were, however, a few exciting landmarks. Beginning with the fall of Rome in 476 and the ascendancy of Christianity about the same time, Europe witnessed such notable events as the reign of Charlemagne (768–814), the Norman conquest of England (1066), the first Crusade (1096), the invention of the magnetic compass (ca. 1200), the invention of the mechanical clock (ca. 1250), and Marco Polo's travels (1270).

The years between 1300 and 1500 witnessed the stirrings of the Renaissance. The invention of printing from movable type by Gutenberg (ca. 1450) and Columbus's discovery of America (1492) heralded the coming of a new day. Without knowing it, Europe stood on the threshold of an era of undreamed-of advances in the sciences, in mathematics, in technology, and in the arts.

Mapmakers and Stargazers

1500–1650

THE influx of Arabic mathematics into Europe gave added impetus to the exploration of the New World as well as to the reception of new ideas. In Portugal, Prince Henry the Navigator exerted tremendous influence in advancing the arts of navigation, mapmaking, astronomy, and shipbuilding. By reason of their extensive commercial activities, the Italian merchant class contributed substantial advances in mathematics.

CHARTING UNTROD PATHS

Navigation beyond the sight of land was possible for early navigators only because they were able to chart their courses by the stars. When the explorers of the New World set forth across the Atlantic, they were able to determine their distance from the equator with reasonable accuracy. Unfortunately, however, they had no satisfactory way of knowing how far westward or eastward they had traveled. For this knowledge, navigators had to wait for the invention of the sextant and other navigational instruments. Thus the need of navigators was one of the compelling motives for devising useful, accurate maps and charts. By the fourth century B.C. the idea that the earth was a sphere had been rather generally accepted among Greek scholars. Aristotle actually proposed six arguments to prove that the earth was spherical. From that time onward the idea was accepted by some scholars, such as the Venerable Bede (ca. 700) and Roger Bacon (ca. 1250), but until the voyages of Columbus and others the prevailing opinion of navigators and geographers was that the earth was flat.

The greatest contributor to geography and cartography in ancient times was Ptolemy (ca. A.D. 140), whose great work on geography consisted of eight volumes. His chief error was in underestimating the size of the earth. The Romans, however, were more interested in military roads and trade routes than in mathematical geography; they fell back on the flat-disk theory. With the fall of Rome, further progress in mapmaking came to a standstill for about a thousand years.

A sudden spurt of interest in mapmaking occurred during the latter half of the fifteenth century. This resulted from two significant events. First, the invention of printing from movable type made possible the wide circulation of Ptolemy's *Geography,* which had been preserved in its original Greek manuscript form. Secondly, the discoveries and explorations of Columbus, Vespucci, Vasco da Gama, Magellan, Cabot, and others added further impetus to the activities of cartographers.

The outstanding mapmakers of the age were Gerhardus Mercator of Flanders, whose real name was Gerhard Kremer, and his friend Abraham Ortelius (Oertel or Ortell), another Flemish geographer. In 1534 Mercator founded a geographical establishment at Louvain, where he began to prepare an elaborate series of maps. At first he was handicapped by the influence of Ptolemy's geography, which had portrayed the Mediterranean several hundred miles too long.

The problem with which Mercator struggled was the same problem that had troubled other cartographers before him: how to represent a spherical surface accurately on a flat sheet of paper. In earlier times, maps limited in scope were relatively accurate and were therefore useful to navigators hugging coastlines. But it was another story when the entire world was to be mapped to help mariners sailing the open seas for thousands of miles. Mercator soon realized that such maps were impossible without considerable distortion and inaccuracy. It is not possible to depict the surface of a sphere on a plane without some kind of misrepresentation. This is because a spherical surface cannot be cut and spread out flat without some pulling or stretching or tearing. Such a difficulty does not exist with a cylinder or a cone, either of which can be appropriately slit open and spread out on a plane without any distortion.

In order to achieve the least harmful distortion, Mercator hit on the idea of "projecting" the spherical surface of the earth onto a cylindrical surface. In other words, imagine a hollow cylinder slipped over the earth, touching it at the equator; a light at the center of the earth could be regarded as casting a shadow of the earth's surface onto the surface of the cylinder. Then, when the cylinder is unwrapped and flattened out, the resulting map is the familiar Mercator projection (1569). On this type of map the meridians of longitude (no longer great circles on a sphere) are vertical, parallel straight lines, spaced evenly apart. Since on the sphere the meridians approach each other and gather together at the poles, it is clear that on the Mercator map, east-west distances are exaggerated more and more the closer one gets to the poles. On the Mercator map the parallels of latitude are horizontal, parallel lines that are spaced farther apart as one gets nearer the poles.

The Mercator map preserves a constant relation between latitudinal and longitudinal distances. This makes it convenient for mariners, who can set their course on a fixed compass direction between two points and follow the line without changing their bearing. The price paid for this advantage is, unfortunately, a distortion of distances, areas, and shapes so serious that for comparing areas it is practically worthless; for example, on a Mercator projection, Greenland and Antarctica appear several times as large as they really are. Mercator's work of a lifetime, however, marks the beginning of the era of modern mapmaking.

Mercator is portrayed in figure 4.1 on the Belgian stamp at the lower right; the portrait of Ortelius, royal geographer to the king of Spain, appears beside it. The map of the world *(top)* is a typical Mercator projection.

Fig. 4.1. Mercator projection (*top*), Ortelius (*l.*), and Mercator (*r.*)

As time went on, other varieties of map projections were devised. Some of them are actual geometric projections; others are constructed by calculations involving advanced mathematics. A "map projection" is a systematic scheme of arranging the earth's meridians and parallels on a flat surface with the deliberate goal of achieving certain desired properties, knowing full well that certain other properties must be sacrificed or at least badly distorted. Thus some projections have equal-area properties—that is, areas on the map correctly measure the areas on the earth that they represent. On other projections, called *conformal* maps, the angle between any two intersecting curves on the map is the same as the angle between the curves on the earth represented by those curves on the map.

In figure 4.2, on the Turkish stamp at the center right, we see an attempt at an equal-area map. The Brazilian stamp *(top right)* shows a so-called homolosine projection devised by J. Paul Goode in 1923, based on an earlier projection by Karl Mollweide. The Cuban stamp shows Mollweide's

Fig. 4.2. Globe; different map projections

homolographic projection of 1805. On the East German stamp we see an earth-globe map from the year 1568, now in the National Mathematics and Physics Collection in Dresden, East Germany.

No discussion of cartography would be complete without mentioning some of the instruments used by navigators and geographers. One of the oldest known astronomical devices was the armillary sphere, used to represent the great circles of the heavens. In its simplest form, the armillary, or celestial sphere, consisted essentially of three rings arranged at right angles to each other and marked off in angular measures, together with a planisphere, which carried a stereographic projection of the heavens on a circular plane surface. Armillaries were first developed by Hipparchus and later improved by Ptolemy. The more elaborate forms of these skeleton globes contained eight or nine rings, including the equator, the horizon, the meridian, the tropics, the polar circles, and an ecliptic hoop.

Armillary spheres were used chiefly by astronomers rather than by navigators. For example: an important ring on an armillary was the one fixed in the plane of the equator; the arrival of the equinoxes was noted by observing when the shadow of the upper half of this ring exactly covered the lower half. In the seventeenth and eighteenth centuries, spherical armillaries were used to demonstrate the difference between the Ptolemaic theory of a central earth and the Copernican theory of a central sun. (In this connection, you may be interested in the article "Maps: Geometry in Geography," by Thomas W. Shilgalis, *Mathematics Teacher* 70 [May 1977]: 400–404.)

Armillaries appear on the three stamps in figure 4.3: the Austrian stamp at the lower right is actually from the title page of a medieval geography book in the map collection of the Austrian National Library and shows Hercules beside the celestial spheres; the Chinese stamp *(top right),* depicting an ornamental armillary of the Ming dynasty, is one of a series of stamps commemorating inventions by ancient and medieval Chinese scientists; and the East German stamp shows an armillary sphere from the year 1687.

The astrolabe, or planisphere, is a more versatile instrument than the armillary and goes back to very early times. A very simple version was the Greek *dioptra* mounted on a graduated circle. As the instrument was improved, it became more complicated but more accurate. Basically, the planisphere consisted of a metal disk with a pointer, or *alidade,* pivoted at its center. It was a portable instrument, ranging in size from five centimeters to sixty centimeters in diameter. When in use, it was suspended from a ring at the top and hung freely of its own weight, so that it remained in a vertical plane. Angles of elevation and depression were easily read from a scale on the rim in conjunction with the pointer. The planisphere could thus be used for surveying as well as navigation. More elaborate forms

Fig. 4.3. Armillaries, or celestial spheres

were equipped with star maps and charts on both sides of the instrument, which served to make it a computer for determining latitude. Astrolabes were frequently ornate and decorative, embellished with astrological and religious symbols. They remained in common use by explorers and navigators even for several centuries after the time of Columbus.

Two astrolabes are seen in figure 4.4: the Iranian stamp on the right, commemorating the 700th anniversary of the death of the Persian astronomer-mathematician, Nasr-ud-Din, depicts an elaborate planisphere type of astrolabe; the stamp from Guinea, commemorating the 500th anniversary of the death of Prince Henry the Navigator, depicts a nautical astrolabe like those used by seventeenth-century mariners. This astrolabe is very similar to the one used by Champlain, which was discovered on the bottom of the Ottowa River in the nineteenth century.

The mariner's astrolabe was constantly used by navigators to determine their latitude by sighting on the polestar (North Star) or the midday sun. Although it was easy to determine the latitude, it was far more difficult to determine longitude. Early navigators therefore often sailed north or south to the desired latitude and then followed that parallel to their destination. This was known as "parallel sailing."

The use of the astrolabe led to the creation of other instruments used chiefly by astronomers, such as the quadrant, the sextant, and the octant.

Fig. 4.4. Nautical astrolabe and planisphere

The quadrant was anticipated by Ptolemy, who described a simple instrument in which a small iron tube was mounted on a quarter-circle cut on a stone block. This was the simplest form of a quadrant; it enabled one to read the angle of elevation of the sun. As time went on, the quadrant developed into a more elaborate form, but basically it was a square or quarter-circle fitted with a plumb line and a movable pointer fitted with sights. Although originally intended and used for astronomical purposes, it evolved into a useful surveying instrument as well. Some quadrants were constructed to revolve in a vertical plane, some in a horizontal plane. In later years rather large quadrants were constructed, sometimes with a radius of as much as six meters.

The Turkish stamp in figure 4.5 *(top right)* pictures a fifteenth-century astronomer's quadrant. The sextant, a relative of the quadrant, was invented by Tycho Brahe in the early seventeenth century. It is shown in figure 4.5 *(bottom)* and on the Australian stamp at the top left, together with a mariner's compass and a picture of the famous Antarctic explorer James Cook.

In modern times, accurate maps have become indispensable for the planning of communities, the development of agricultural and industrial areas, and the exploration of natural resources. Mapmaking concerns the surveyor determining the boundaries of real estate, townships, counties, and so on; it involves the engineer laying out a highway system or an irrigation scheme; it affects the prospector of a potential oil field or a possible source of uranium. The day is past when the compass, the theodolite, and the surveyor's chain alone were adequate for producing maps. Today the mapmaker uses aerial photography, automatic stereoplotters, electronic

Fig. 4.5. Sextants and a quadrant

distance measuring, airborne profile-recording instruments, and laser terrain-profiling equipment. But we are getting slightly ahead of our story.

SCANNING THE SKIES

The close of the fifteenth century and the early part of the sixteenth century was truly an Age of Discovery. Not only did navigators and explorers open up a new world of continents, but scientists such as Copernicus, Kepler, and Galileo had the courage to question long-established doctrines and to open up a new world of ideas. The Greek astronomy of Hipparchus and Ptolemy had come to a dead end. According to that system, all the heavenly bodies were supposed to revolve around the earth, which was at the center. The system worked fairly well when it came to predicting the positions of the planets, but Copernicus objected to its complications. It was the destiny of Copernicus to revolutionize the popular concept of the universe.

Well over one hundred stamps have been issued in honor of Copernicus. The choice is an embarrassment of riches; almost any other three would have been as satisfying as the three shown in figure 4.6. The Chinese stamp *(top left)* is intriguing for its simplicity; the other two were both issued to commemorate the 500th anniversary of the birth of Copernicus.

Fig. 4.6. Copernicus

Born in Torun, Poland, in 1473, Copernicus spent some years in Italy, where as a young man he studied law and medicine and became interested in astronomy. As he thought about the planets, he felt that their positions could be calculated more simply and more accurately if he supposed that the sun were stationary and that the planets, including the earth, all revolved around the sun instead of around the earth. This was not a new idea; it had been suggested earlier by Cusanus and still earlier by the Greek, Aristarchus of Samos. But Copernicus did not stop there. Although he made few observations and fewer measurements, he worked out a complete theory, mathematically, to show how it was possible, on this assumption, to calculate the positions of the planets. Even though he kept the circular orbits of his predecessors, his theory explained the facts better than the old system. Having worked it out in detail, he wrote a book about it, but he was reluctant to publish it for fear of offending the Church. Eventually he allowed it to be published, dedicating it to Pope Paul III. But he was cautious enough to emphasize that his description of the universe was merely a *theoretical scheme* for calculating planetary positions rather than an actual description of the real universe. Thus he was spared the agony suffered a few years later by Galileo, who was persecuted for holding a similar view. And so began the revolution that overthrew Greek astronomy forever.

Copernicus died in 1543, three years before Tycho Brahe was born. Where Copernicus was a theoretician, Brahe was a superb observer. Copernicus had built his scheme on a narrow basis of facts. If the earth also rotated on its axis, as suggested, there were some facts that the Copernican theory did not explain. More accurate observational data were needed, and these were carefully and patiently collected by Brahe.

Tycho Brahe was a Danish nobleman of considerable means. As a young man, he was lucky to have discovered a *nova,* or new exploding star, and was soon recognized as a skillful astronomer. With the generous support of King Frederick II of Denmark, he established an elaborate observatory on the tiny island of Ven, some twenty miles north of Denmark. He called the observatory Uraniborg, or "Castle of the Heaven." It was lavishly equipped at royal expense with all the necessary astronomical equipment. Anxious to obtain the greatest possible accuracy in his observations, Brahe designed and built unusually large instruments, and for more than twenty years he labored assiduously, gathering a vast number of observations of unprecedented accuracy.

Toward the end of his career, Tycho Brahe fell into bad grace with the new king, Christian IV, and was deprived of his great observatory. Thereupon he left for Germany, where by good fortune he found a young assistant by the name of Johannes Kepler (1571–1630), to whom he turned over all his patiently accumulated observations. From this mass of data Kepler prepared tables of planetary motions, continuing the work of Brahe, who died a few years later. Shortly thereafter, the magnificent equipment at Uraniborg became useless, for Galileo's telescope had supplanted the naked-eye instruments forever. In figure 4.7, the Austrian stamp on the left

Fig. 4.7. Johannes Kepler and Tycho Brahe

has an excellent portrait of Kepler, and the Danish stamp on the right portrays Tycho Brahe.

The German astronomer Kepler was beset from early childhood by poverty and ill health. As a young man he began to study for the ministry, but his mathematical ability was soon recognized, and he became deeply involved in astronomy. His burning curiosity concerned the *motions* of the heavenly bodies as well as their sizes and positions. Like Copernicus, Kepler was a mathematician rather than an observer. But he appreciated the superb observations that he had inherited from Brahe and spent the next twenty-five years working on them. His object was to devise a scheme, based on this invaluable data, that would explain planetary orbits more satisfactorily than the theories of his predecessors. Unfortunately, he was misguided, probably because he was sympathetic to the ancient Pythagorean doctrine of the "music of the spheres." At first he was unsuccessful, but he did not give up easily. Deciding that circular orbits just didn't work, he began to look for some other kind of curve. Somehow he hit on an elliptic path and was much encouraged when he found that the positions of Mars, as observed by Brahe, fitted into an elliptic orbit with remarkable accuracy. On further study, he found that the orbits of the other planets also followed elliptic paths, with the sun always at one of the focal points of the ellipses. This remarkable result he announced as the first of three laws of planetary motion. These "laws" were entirely empirical, that is, they were suggested solely by actual measurements and not by any theoretical considerations. Kepler's theory of elliptical orbits struck the final deathblow to Greek astronomy, and his explanation of the solar system has been followed by astronomers to this day. Nevertheless, a number of questions remained unanswered in Kepler's day: Why was the sun always at one focus of the ellipse? Why did the heavenly bodies always remain in their orbits? These and other questions were not to be answered until the time of Newton, half a century later.

We have seen that Copernicus and Kepler were mainly theoretical mathematicians, whereas Tycho Brahe was basically an observer. Galileo (1564–1642), however, although not primarily a mathematician, was able to combine experimental and observational techniques with an analytical mind and a keen perception of physical laws. He was a man of many talents. His father wanted him to become a physician; however, he was eventually persuaded to let his son study science and mathematics. Galileo had an unusual ability to generalize and to seek overall, governing principles. Moreover, he had a literary gift for expressing himself clearly and forcefully.

Although many stamps honor Galileo, the two in figure 4.8 portray him as he is usually shown. Both stamps commemorate the 400th anniversary of his birth. Galileo is remembered not only for his contributions to astronomy but also for his discoveries in physics and dynamics. His interest in the mechanics of motion was first aroused by observing a swinging lamp in a cathedral, from which he deduced that the time of a complete swing was independent of the weight of the swinging object. From this he was led to study the behavior of freely falling bodies. His reputed experiment of dropping two heavy cannon balls of different weights from the Leaning Tower of Pisa may be somewhat legendary, but his conclusion was sound, namely, that the time of falling was independent of the weight, provided air friction did not interfere.

Fig. 4.8. Galileo

Galileo was a keen observer who insisted on careful measurements. But when it came to measuring small intervals of time, he worked under a handicap. Adequate clocks had not yet been perfected, and he had to measure time by using his pulse or by noting the rate at which water trickled through a small hole and was caught in a container. Yet he was able to clarify the concept of uniformly accelerated motion (using an inclined plane in his experiments) and to arrive at two basic principles: (1) that the speed of a falling body increases steadily as time passes, and (2) that the distance fallen is proportional to the square of the time elapsed. These principles were in direct contradiction to the views held by Aristotle.

In arriving at his conclusions, Galileo put his mathematical insight to good use, even though he employed geometric methods in his reasoning, for the tools of analytic geometry and the calculus were not yet available. He believed firmly in the necessity for applying mathematics to the examination of physical phenomena. Galileo dealt with many aspects of physics: he devised a crude gas thermometer; he showed that a body could be acted on by two or more forces at the same time; he clarified the motion of projectiles and improved the science of gunnery; he recognized that water could not be pumped to more than a certain height; he stated that when a flexible chain carrying only its own weight is suspended from two points, the curve formed is not a parabola; he studied the strength of materials, showing the relation between the volume and strength and explaining why, for example, a hollow tube may be stronger than a solid rod of the same dimensions and material; he even suspected that light had a finite velocity but was unable to measure it.

Although he agreed with Copernicus that the earth revolved around the sun, Galileo was at first cautious not to say so in public. Having heard that a "magnifying tube" using lenses had been "invented" in Holland, Galileo proceeded to construct a telescope of his own design. With it he discovered many undreamed-of things in the sky: the mountains on the moon, the spots on the sun, the four satellites of Jupiter, and many new stars. (The Albanian stamps in fig. 4.9 show several planets in orbit, in particular, Venus, with a rocket, and Saturn with its rings.)

Fig. 4.9. Venus and Saturn

When Galileo first announced these astonishing discoveries, there was great enthusiasm but also much disbelief and anger, especially on the part of the Catholic church. The very thought of "invisible" objects in the heavens and the idea of the earth dethroned from its exalted position as the center of the universe infuriated and outraged leaders of the Church. The pressure soon became so great that the Church denied the doctrine of Copernicus and proclaimed Galileo a heretic.

For a while Galileo, fearful, dared say no more. The time came, however, when he felt that he had to speak out, which he did by courageously publishing in 1632 a book called *A Dialogue on the Two Chief World Systems*. It was a subtle literary device in which two characters argue before a third person. One presents the Ptolemaic doctrine; the other argues for the Copernican viewpoint. Naturally, the expounder of the Copernican theory gets the better of the argument. But the Church saw through this subterfuge. Galileo was brought before the Inquisition, tried, found guilty of heresy, forced to renounce his theory publicly, and imprisoned. After years of persecution, the Church relented slightly, allowing him to retire to a villa near Florence. Nearing the age of seventy, ill, and nearly blind, Galileo still had the courage of his convictions despite outward appearances. The world of scholarship knew that he was right. The revolution in science that had begun a century earlier with Copernicus had fully arrived.

ASTRONOMICAL INSTRUMENTS

We have already described the armillary sphere, the astrolabe, the sextant, and the quadrant. During the seventeenth century armillary models were frequently used to explain the difference between the Ptolemaic and the Copernican theories.

As for Galileo's contribution to astronomy, much credit must be given to the telescope, which opened up a new world. Glass was known to the Egyptians as early as 3500 B.C., and crude lenses have been discovered dating from 2000 B.C. The Arab scientist Alhazen (discussed in chapter 3) experimented with magnifying lenses and parabolic mirrors in the tenth and eleventh centuries. In 1608 in Holland, Hans Lippershey built a number of telescopes to be used for military purposes. (He did not invent the telescope; such instruments were available in France and Germany at the time.) Galileo reinvented and improved on the device, grinding his own lenses. His largest telescope was about 4.5 centimeters in diameter and had a magnifying power of thirty-three diameters.

Several types of telescopes have been shown on stamps. The Russian stamp in figure 4.10 *(center)* commemorates the Tenth Congress of the International Astronomical Union in Moscow and shows a modern conventional telescope. The Japanese stamp *(left)*, commemorating the fiftieth anniversary of the Mizusawa Latitudinal Observatory, shows a floating

Fig. 4.10. Telescopes

zenith telescope. The East German stamp on the right shows a modern Zeiss telescope with a galaxy in the background.

Optical telescopes have two important characteristics: their magnifying power and their light-gathering capacity. Few scientific instruments have had such a dramatic impact on our culture as the telescope. From the seventeenth century on, telescopes were continually improved, increasing in size and power. There are two basic types of optical telescopes: reflecting and refracting.

In a reflecting telescope, the light is gathered and focused by a mirror. The large telescopes are reflectors; for example, the largest of this type, built in the early 1970s in Leningrad and installed in the Caucasus, weighs 765 metric tons and has a 25-meter tube, a 400-centimeter diameter reflector, and a mirror nearly 6 meters in diameter.

In a reflecting telescope (such as that used by Galileo) the large lens or object glass is convex and the eyepiece, placed in *front* of the focus, is concave; the resulting vertical image is upright. With such an instrument (like ordinary opera glasses), both the magnification and the field of view are limited. In the modern visual refracting telescope, however, the eyepiece is a convex lens placed in *back* of the focus, resulting in an inverted image, which can be reinverted. The result is increased magnification.

Over the years, both types of telescope have been developed to a high degree of perfection. Each type has its own advantages and disadvantages and serves its special purposes. Both types are in use by contemporary astronomers.

The modern observatory gradually evolved along with the telescope. One of the earlier observatories is shown on the Korean stamp in figure 4.11. The Russian stamp at the upper left shows the Crimean Observatory; the U.S. stamp at the lower right shows the Palomar Mountain Observatory in California, dedicated in 1948. The Japanese stamp, commemorating the seventy-fifth anniversary of the Tokyo Astronomical Observatory, shows the telescope through the revolving open segments. The movable segments of the conventional observatory roof dome were invented by Maximilian Hell (1720–1790), a Slovakian Jesuit and astronomer.

Since the early 1930s, the radio telescope has come into prominence. Basically, it depends on the fact that a parabolic "dish" can pick up radio wavelengths from outer space. Some very large radio telescopes have been built. The Jodrell Bank radio telescope near Manchester, England, for example, was one of the first to be built. It is a remarkable piece of engineering: the paraboloid dish is some seventy-five meters in diameter. The use of radio telescopes in connection with high-speed electronic computers has yielded invaluable data to modern astronomers. It must be said, however, that the radio telescope has not supplanted the optical telescope. Each has its own specific uses; both play a vital role in today's astronomy.

Fig. 4.11. Observatories

On the one hand, the radio telescope is well adapted to tracking satellites in orbit, which the optical telescope cannot do. On the other hand, the radio telescope cannot yield photographs of stars and galaxies.

Some of these radio telescopes can be seen in figure 4.12. The Israeli stamp at the upper left shows a parabolic receiver at a satellite-tracking station; the British stamp *(upper right)* depicts the Jodrell Bank radio telescope; the French stamp shows the huge radio telescope at Nançay; and the Haitian stamp shows a radio telescope at a modern observatory.

Having prematurely caught a glimpse of the twentieth century, we take our leave of the mapmakers and stargazers of the sixteenth and early seventeenth centuries. This period of ferment witnessed many notable events. In 1521, Martin Luther was excommunicated; in 1534, the Jesuit order was founded by Ignatius Loyola. About 1550 Michelangelo was at the height of his fame. Queen Elizabeth I ascended the throne of England in 1558, and around 1600 the age of Shakespeare was in full flower. Half a century later, in 1654, Louis XIV was crowned king of France.

Fig. 4.12. Radio telescopes

The advent of the Industrial Revolution was also clearly evident during these years. Among a few of the significant advances were the iron-rolling machine (1552), the screw lathe (1578), the compound microscope (1590), the use of gunpowder in mining (1613), the submarine (1624), the fountain pen (1636). Francis Bacon, in his *Novum Organum* (1620), advocated the scientific method, stressing inductive reasoning for science rather than the deductive reasoning as used in mathematics. It was the time of a great awakening. The world stood on the threshold of the birth of modern mathematics.

The Awakening
1650–1800

BURSTING with new ideas, the next century and a half was an exciting period that ushered in the age of modern science. Galileo had annihilated the Ptolemaic system, Kepler had discovered the laws of planetary motion, and Newton was to lay the foundations of classical mechanics. Aristotelian science was buried, and the scientific method was in the ascendancy. It was also a period that marked the first, mild industrial revolution. The waterwheel replaced human power; cheap paper became available; glass came to be used in construction. The telescope, the thermometer, and the pendulum clock were invented. The stage was being set for the second industrial revolution.

The mid–seventeenth century saw the birth of modern mathematics: (1) the analytic geometry of Fermat and Descartes, in which the two concepts of *number* and *form* were fruitfully joined; (2) the differential and integral calculus, created by Newton and Leibniz; (3) the theory of probability, explored by Fermat and Pascal; and (4) the dynamics of Galileo and the universal gravitation of Newton. That modern mathematics should have blossomed at the height of the baroque period was not primarily because of a clamor for "practicality" or the utilitarian promise of science. Mathematics flowered because of the dreams of men like Fermat and Descartes, Newton and Leibniz. In the main, their ivory-tower thoughts were far removed from practical applications.

NATURAL PHILOSOPHERS

Before describing these significant developments, let us look at the activities of some natural philosophers, notably Stevin, Torricelli, and von Guericke. Simon Stevin (1548–1620), seen on the Belgian stamp at the left in figure 5.1, was a mathematician of considerable versatility. Born in Belgium, he lived most of his life in the Netherlands, serving variously as inspector of dikes, quartermaster in the Dutch army, and minister of finance; in the latter position he introduced Italian methods of bookkeeping and accounting into Holland. His chief contribution was popularizing the theory and use of decimal fractions, which he did by publishing in 1585 a book called *La disme.* Practical merchants and engineers quickly recognized the convenience of the new notation and readily adopted it, but it was not until about a hundred years later that decimal fractions were accepted by the general public. In his enthusiasm for the decimal idea, Stevin even suggested a decimal division of degrees of angles as well as a decimal subdivision of coinage. Today, decimal division is commonly employed even where the metric system is not yet in use—for example, decimal divisions of the inch in machine-shop work and of the foot in land surveys. These established practices are in no way related to the metric system.

Fig. 5.1. Stevin, Torricelli, and von Guericke

Equally significant were his contributions to science, notably in statics and hydrostatics, where he enunciated three important principles: (1) by means of a frictionless, endless chain draped over inclined planes, he showed that perpetual motion was impossible; (2) by dropping two different weights simultaneously from the same height, he noted that they reached the ground at the same time, as Galileo had already found; and (3) he proved that the pressure exerted by a liquid depends on the area of the surface and on the height of the liquid above that surface but is independent of the shape of the container. Stevin wrote in Dutch, which marked the beginning of the end of the custom of European scholars writing in Latin. The change by other scientists to writing in their native language was very gradual, and even a century later British mathematicians were still using Latin.

Other notable scientists include the Italian, Evangelista Torricelli, seen on the Russian stamp in the center in figure 5.1, and Otto von Guericke, seen on the West German stamp on the right.

Having studied mathematics in Rome, Torricelli (1608–1647) became an admirer of Galileo, whom he had come to know during the last few months of the aging man's life. Galileo had wondered why, when water was lifted by a pump, it would never rise beyond a certain height. Torricelli guessed that air had weight and that it was this weight that pushed against the water. He guessed further that air weighed only enough to balance a column of water about 10.3 meters high. To verify this guess, he devised a famous experiment. Using a glass tube open at one end, he filled it with mercury and inverted it so that the open end dipped below the surface of some more mercury in a dish. The tube was about 122 centimeters long; when he inverted it as described, the mercury in the tube dropped to a height of about 760 millimeters and remained there, held up by the weight of the outside air pressing on the mercury in the dish. The space in the tube above the column of mercury formed a vacuum—the first made by human hands. Noticing that the height of the mercury column varied slightly from day to day, Torricelli assumed that the atmospheric pressure changed as weather conditions changed. And so the barometer was born. That is why the National Weather Service refers to so many "inches of mercury."

The German physicist von Guericke (1602–1686), a contemporary of Torricelli, also became interested in this matter of a vacuum, which Aristotle had insisted could not exist. To find out, Guericke independently devised many experiments, using a crude air pump that he fashioned himself. It was an unwieldy and inefficient pump, but it worked. His most dramatic experiment was with two large metal hemispheres carefully fitted together with a layer of grease between their edges. These were the famous Madgeburg hemispheres, shown on the East German stamp in figure 5.2. Placing the two hemispheres together, he exhausted the air by means of the pump; then, to the astonishment of the spectators, two teams of horses pulling on each hemisphere were unable to pull them apart. But when air was allowed to reenter the joined hemispheres, they fell apart readily. When von Guericke heard of Torricelli's experiments, he realized that it was the weight of the air outside the hemispheres that held them together, not some mysterious force called a vacuum.

Fig. 5.2. Magdeburg hemispheres

The significance of these experiments with vacuums should not be underestimated. In time, further experimentation on the subject led to the invention of the steam engine, which in turn had a tremendous impact on the Industrial Revolution then beginning. The subsequent development of gas engines, steam turbines, and jet-propelled planes are familiar enough in our present-day culture.

One other scientist who made significant advances in astronomy and physics about this time (ca. 1660) was Christian Huygens (1629–1695), who appears on the Dutch stamp at the right in figure 5.3. Educated in mathematics, the young scholar turned his attention to science instead. In astronomy, his efforts were devoted to improving the utility of the telescope, which he did in several ways. For one thing, he made better lenses, grinding them himself. To make telescopic measurements more precise, he designed a micrometer for making better angle measures. He invented the pendulum clock to measure time more precisely; heretofore the only clocks available were water clocks and mechanical clocks operated by gears

Fig. 5.3. Huygens's pendulum clock, mechanical clock, and Christian Huygens

and weights. The clock shown on the Austrian stamp (*center,* fig. 5.3) has a mechanical gear clockwork such as that used in seventeenth-century clock towers. The picture of Huygens's pendulum clock in the Dutch stamp at the left is taken from a painting by van Ceulen. Huygens adapted Galileo's principle of the pendulum to the clock mechanism, using suspended weights on pulleys, which allowed the pendulum to swing without stopping for a definite period. This was an invaluable advance for physics, since measuring time accurately is indispensable to that branch of science.

Huygens also devoted careful study to the properties of light. He favored a wave theory of light despite the fact that contemporary thought held to the theory that light consisted of extremely small particles (corpuscular theory of light).

THE BIRTH OF MODERN MATHEMATICS

The age in which René Descartes lived was beset by wars and religious bigotry, albeit signs of transition to an intellectually brighter future were appearing. Born in 1596 at La Haye, France, near Tours, Descartes had fragile health as a child. He received the conventional Jesuit schooling of his day—languages, mathematics, metaphysics, history, logic, rhetoric, and ethics. As a young man he decided to find out for himself how people think. All his life Descartes was a loner, seeking solitude by incessantly roaming from place to place and spending years in the anonymity of military service. Although a perceptive observer, his antisocial nature prevented his active participation in social life, and he became a recluse, devoting his energy to scholarship.

Primarily a philosopher, Descartes was also a physicist, mathematician, and a pioneer in physiology. All his efforts were dominated by his concern for logic and methodology. Dissatisfied with conflicting contemporary philosophies, Descartes would accept nothing unless it was so clear to his own mind that he could not doubt its truth. To him, exact knowledge could be arrived at only by rigorous logical reasoning. This principle applied to science as well as to mathematics, although he did not rule out experimentation. He felt that mathematics was the only certainty. He further believed that the cause and effect of every event or phenomenon could be known by the methods of logic and mathematics. If applied to the physical environment, all the secrets of nature could be discovered, and the environment could be manipulated and controlled.

Descartes regarded mathematics as having external objective existence; that is, angles, triangles, and so on, exist whether we think of them or not. Thus he leaned heavily on intuition for mathematical concepts and asserted that logical deductions made from these intuitive ideas must be truths that apply to the physical world. Today this sounds familiar enough, but at the time it was a novel idea. Descartes is shown on the French stamp in figure 5.4.

Fig. 5.4. René Descartes

In 1637 he published his celebrated *Discours de la méthode pour bien conduire sa raison et chercher la vérité dans les sciences* ("Discourse on the Method of Reasoning Well and Seeking Truth in the Sciences"). This revolutionary book brought him fame overnight, for it was indeed a breakthrough in thought. *La géométrie,* an essay of a hundred pages, was actually one of several appendixes to the *Discours,* which was devoted to vortices and the solar system. In *La géométrie,* Descartes exemplified the power of his methodological approach. He was not happy with the methods of proof used by the ancient Greeks, and he also felt that the algebra then current was lacking in *general* methods. His creative achievement lay in the ingenious way in which he applied algebra to geometric problems. Although Descartes did not use coordinate axes and ordered pairs (or triples) of numbers as we do today, he realized the superior power of general algebraic methods over the "synthetic" methods used by the Greeks. The use of algebraic equations to represent geometric curves and to study their properties did not catch on at once, and it was several decades before "analytic" geometry was widely accepted. It paved the way for the development of the calculus, which depends on curves to explain physical phenomena like velocity and acceleration.

In the synthetic geometry of the Greeks, points and lines are not defined logically, only intuitively, whereas in analytic geometry points and lines are defined algebraically (i.e., analytically). Thus a point is defined as an ordered pair (or triple) of numbers, and a line is defined as an equation of the first degree. The superiority of the analytical method over the synthetic method can scarcely be exaggerated.

The contributions made by Descartes to philosophy and to science were surpassed only by his invention of analytic geometry. His philosophy was heretical. Had it not been for the encouragement of Cardinal Richelieu, Descartes might never have published the *Discours de la méthode,* for fear of persecution by the Church.

His life was brought to an untimely end at the age of fifty-four as a result of pneumonia contracted while he was in the service of Queen Christina of Sweden; she had invited him there some few months earlier to instruct her in philosophy.

Someone has written that Blaise Pascal (1623–1662), a contemporary of Descartes, achieved a sevenfold immortality: as a philosopher, a mathematician, a physicist, a religious zealot, a saint, a mystic, and a scoundrel. Be that as it may, the fact remains that he was a child prodigy, who at the age of sixteen wrote a pamphlet on the conic sections, at eighteen invented a calculating machine, and at twenty-four discovered properties of fluids. Pascal is pictured on the two French stamps in figure 5.5.

Fig. 5.5. Blaise Pascal

His book on the geometry of conic sections was a brilliant piece of work that carried the subject far beyond the work of Apollonius. In this connection, Pascal, together with another scientist, Desargues, took over ideas from the theory of perspective suggested by Leonardo da Vinci and Albrecht Dürer. As a result, the formal science of projective geometry was born. Unfortunately, it was overshadowed by the then universal interest taken in the newly created analytic geometry of Descartes, and so the significance of projective geometry was not recognized until some two hundred years later.

The calculating machine that Pascal built, using cogwheels, could add and subtract, but instead of achieving fame and fortune as he had hoped, Pascal found the venture unprofitable. Becoming involved in religion, he was persuaded to Jansenism, a Catholic sect opposed to the Jesuits. His zeal was amazing, for up to then he had embraced science and reasoning as a way of life. These he now forsook in his enthusiasm for religion. But this zeal was short-lived, for he soon turned his attention to physics. The question of a vacuum was still a lively one, and Pascal engaged in experiments similar to those of Torricelli's. Using many different tubes and a variety of liquids, he reaffirmed the nature of atmospheric pressure, even proving that a barometer would register a difference in height when taken to the top of a mountain. His experiments also led to the discovery of the principle of the hydraulic press and other physical properties of liquids.

In his late twenties, because of personal problems, Pascal changed his mode of life, becoming somewhat more urbane and socially active, or at least trying to live a normal life without being unduly preoccupied by piety. But in 1654, at the age of thirty-one, his fortunate escape from what could have been a fatal accident caused him to seek the solace of religion once more; this time he turned his back for good on a worldly life.

It was at this time that he began corresponding with the lawyer-mathematician Pierre de Fermat on the subject of probability. It all began when a prominent gambler of the time posed a problem to Pascal. The problem related to the division of stakes between two players who quit before their game was to end. As a result of their collaboration, Pascal and Fermat between them laid the foundations of the modern mathematical theory of probability. The theory that was initiated by a gambler's problem eventually was to influence profoundly not only statistical applications in biology, economics, education, and other fields but also—and perhaps even more significantly—today's conception of the physical universe, as suggested by the relativity theory and quantum mechanics. As we shall see later, today's physicists do not think of an electron as being "at" a given place at a given instant of time; instead, they compute the probability of its being somewhere in a given region.

Toward the end of his brief career, Pascal affirmed that he could not reconcile science and religion; he deemed religion far more important than science and became associated with a convent near Paris. There he spent his last years, refusing in his bitterness to recognize Descartes's creation of analytic geometry. His final illness brought on an untimely death at the age of thirty-nine.

Isaac Newton (1642–1727) was a weak, sickly child who delighted in creating all sorts of mechanical devices and in conducting various experiments. He showed little aptitude for academic pursuits until he was eighteen, when he entered Cambridge University, graduating five years later, in 1665. About this time he made some startling optical experiments with light and glass prisms, which made him famous overnight. In 1669 his mathematics teacher, Isaac Barrow, resigned his professorship in favor of Newton, who thus found himself a professor of mathematics at Cambridge at the age of twenty-seven.

Three portraits of Newton are shown in figure 5.6 on stamps from France, Poland, and Mexico.

Fig. 5.6. Isaac Newton

Newton made three memorable contributions to science and mathematics: (1) in physics, his corpuscular theory of light; (2) in astronomy, the theory of gravitation; and (3) in mathematics, the calculus.

The experiments with prisms and color led Newton to speculate on the nature of light. Since light rays move in straight lines and throw sharp shadows, Newton concluded that light rays consisted of small particles—the corpuscular theory of light. However, other scientists of the time, notably Huygens and Robert Hooke, believed that light, like sound, was transmitted with a wavelike, periodic motion. Because of Newton's prestige, the particle theory was favored over the wave theory for nearly a century. Today physicists accept both, explaining some phenomena by one theory, others by the other theory. Because of his convictions about the nature of light, Newton devised the reflecting telescope, a great improvement over the earlier refracting telescope.

In the early 1680s, Newton was challenged by the discussions of Hooke, Halley, and Wren on the motions of heavenly bodies. As a result, he came up with his theory of gravitation, which was set forth in his famous book, *Philosophiae Naturalis Principia Mathematica* ("Mathematical Principles of Natural Philosophy"). Written in Latin and published in 1687, it has been called one of the greatest scientific books ever written. Here were stated the three famous laws of motion:

1. *If a body is at rest, it remains at rest unless acted on by outside forces; a body in motion remains in motion at a constant velocity unless acted on by outside forces.* This is the "law of inertia."
2. *A change in motion is proportional to the force causing the change in the direction in which the force is acting.* This law distinguishes the *mass of a body* (i.e., its resistance to acceleration) from the *weight* of the body (i.e., the amount of gravitational force exerted on it by another body, usually the earth).
3. *To every action there is always an equal and opposite reaction.* (A book lying on a table presses down on the table, and the table pushes up on the book; a bat hits a baseball, and the ball pushes back on the bat.)

These laws of motion applied not only to phenomena on the earth but also to the motions of celestial bodies. Thus the forces attracting the moon and the earth, or any other planets, follow a universal law of gravitation. Expressed mathematically, this law is written

$$F = \frac{km_1m_2}{d^2},$$

which simply means that the force of attraction F between any two bodies is directly proportional to the product of their masses m_1 and m_2 and inversely proportional to the square of the distance d between them. This Newtonian, or classical, physics was accepted as the final word for the next three hundred years until Einstein's theory of relativity was accepted. Einstein's theory does not contradict the Newonian theory; rather, it includes it as a "special case."

Most of us think of Newton as the "inventor" of that unique branch of mathematics known as the calculus. Strangely enough, the calculus was developed independently by both Newton and Leibniz at about the same time. At first the two men were friendly, but as time went on and each became more famous, the controversy over who invented it first became disagreeable and was colored by nationalistic emotions: who deserved the credit, England or Germany? Ironically, as is often true of simultaneous independent inventions, the time was ripe. The idea was in the air. Moreover, the necessary clues had already been anticipated by others: Fermat, Huygens, Wallis, and Barrow.

Although the two mathematicians approached the subject in somewhat different ways, their basic objectives were the same. They were seeking general methods of finding maximum and minimum values of a curve, of constructing tangents to a curve, and so on. The details are too technical to go into here. It is sufficient to say that both men were superb mathematicians. Newton, unhappily, devised a rather clumsy notation, whereas Leibniz, in creating the dy/dx notation to indicate a rate of change, inadvertently fashioned a far more convenient tool. Indeed, the obstinacy with which British mathematicians clung to Newton's notation and geometric methods held back the progress of mathematics in England for nearly a century.

Newton was perhaps one of the greatest intellects the world has ever known. Throughout his entire career he was accorded the highest respect ever shown any scientist, with the possible exceptions of Archimedes, Gauss, and Einstein. Yet he was very human. His were the minor weaknesses often found in the great: he was shy, disliking controversy and publicity; he was extremely sensitive to criticism. He was rather contentious and unbelievably absentminded. He never married.

Gottfried Wilhelm von Leibniz (1646–1716) appears in figure 5.7 on a Romanian and two German stamps.

Leibniz thought of quantities as becoming small enough to be neglected; Newton's "fluxions" involved essentially the same concept. Both men were

Fig. 5.7. Gottfried Wilhelm von Leibniz

groping for what we today call the theory of limits. It would probably be fair to say that Leibniz was inferior to Newton both as an analyst and as a geometer. Nevertheless, he was a creative and philosophical mathematician with a flair for mathematical form and a penchant for finite solutions rather than solutions in infinite series.

Leibniz was a child prodigy; as a mature man he revealed a remarkable range of interests—diplomacy, politics, economics, international law, science, philology, and philosophy. In this latter role, his search for a "universal characteristic" anticipated what is known today as symbolic logic. He also devised a calculating machine that was an improvement over Pascal's machine, being capable of not only adding and subtracting but multiplying, dividing, and extracting roots.

MATHEMATICAL ANALYSIS

The method of the calculus originated largely as an analysis of the changing motion of a moving object. Such an analysis was inconceivable to the minds of the ancient Greeks, who sought only what was static and unchanging. To them, since the circle was beautifully perfect, perfection was a world in which all motion was uniform, following a circular path.

Thus the invention of the calculus marked the beginning of modern mathematics as distinct from all that had preceded it. Few other single developments in the history of mathematics have borne such a rich harvest. For the next 150 years mathematicians exploited this powerful new tool, applying it to many kinds of problems in mechanics, astronomy, and the sciences. Mathematicians such as the Bernoullis, Euler, Lagrange, Laplace, and Gauss worked prodigiously in exploring and refining the methods of the calculus.

As already suggested, the superiority of the Leibniz notation helped to spread the new mathematical method throughout Europe, where it gave rise to mathematical analysis. Among the first to encourage this development were the Swiss brothers, Jacques and Jean Bernoulli. Theirs was an amazing family; in all, about a dozen Bernoullis, over some four generations, distinguished themselves as mathematicians. Not one of them has yet been honored on a stamp.

One of the most brilliant devotees of the methods of Newton and Leibniz was another Swiss mathematician, Leonhard Euler (1707–1783). The father of modern mathematical analysis, Euler did for the calculus what Euclid had done for the geometry of his predecessors by placing it on a broader and more inclusive basis. At the same time, he introduced new symbols and terminology. Much of the notation used in today's mathematics is due to Euler's influence.

Euler is portrayed in figure 5.8 on the three stamps from the German

Fig. 5.8. Leonhard Euler

Democratic Republic, Switzerland, and the USSR. The Swiss stamp *(upper right)* bears the equation $e^{i\phi} = \cos \phi + i \sin \phi$ along the left edge, a fundamental equation of modern analysis, first introduced by Euler.

At the age of twenty-six, Euler married and settled down to professional work in earnest. He had just been made chief mathematician of the recently established Saint Petersburg Academy. Before he was thirty, he lost the sight of one eye, but this did not deter him. He was a prolific writer, publishing more than five hundred books and papers during his lifetime. Original ideas seemed to flow effortlessly from his pen. He had remarkable powers of concentration and could carry on extensive computations mentally. He was able to plan and think through a new idea to completion first and then write it down in final form at once. Although his native tongue was German, Euler generally wrote in Latin and often in French.

At the age of thirty-four he was invited by Frederick the Great of Prussia to join the Berlin Academy, but his brief stay there was not happy. At the behest of Catherine the Great, he returned to Russia in 1776. Some few years later he lost the sight of his other eye, and the last fifeen years of his life were spent in total darkness. Undeterred, still he continued to write, producing new research papers until his death at the age of seventy-six.

Euler contributed to virtually every branch of pure and applied mathematics. In astronomy, he made a special study of the motion of the moon, substituting analytical methods for the geometric methods used by Galileo and Newton. In physics, he subscribed to the wave theory of light and believed that color was due to wavelength, a conjecture that was later proved correct.

A great organizer of mathematics, he interrelated the different branches of the subject, filling in missing gaps. He treated everything analytically; thus trigonometry, which for centuries had been essentially geometric in nature, literally became a new subject when he applied algebra to it. He developed the theory of logarithms and exponents, inventing the symbol e to represent the limit of the exponential series:

$$e = (1 + 1 + \frac{1}{2!} + \frac{1}{3!} + \frac{1}{4!} + \ldots) \approx 2.718\ldots$$

In this connection he arrived at analytical formulas involving trigonometric functions and complex numbers. One of the most remarkable of these relations is

$$e^{i\pi} + 1 = 0,$$

an identity involving five basic constants of mathematics. This can be derived from a more fundamental relation, also due to Euler, that

$$e^{i\theta} = \cos\theta + i\sin\theta,$$

where i represents the imaginary number $\sqrt{-1}$; simply let $\theta = \pi$, where $\pi = 180°$. He contributed to the theory of numbers and also to the nature of infinite series, inventing the standard symbol Σ for "summation." He developed the theory of Latin squares, which at the time was suggested by a recreational problem but which today is used in statistics when planning experiments involving several variables.

Another famous recreational problem was that of the bridges of Königsberg. The city, divided by a river, included two islands connected by seven bridges to the mainland as shown in figure 5.9. The problem was, Could a man leave home, take a walk, and return home, crossing each bridge only

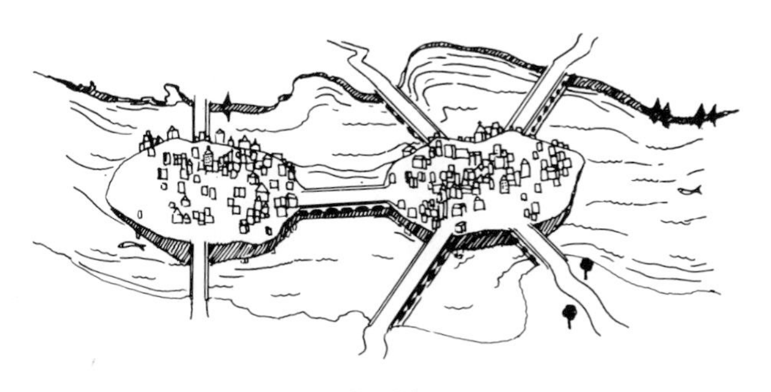

Fig. 5.9

once? In showing why this was mathematically impossible, Euler created a new branch of mathematics, the theory of networks, or graph theory, as it is also known. Today graph theory is indispensable in many problems of technology and industry.

Although progress in mathematics in England had been held back by a persistent adherence to Newton's notation, on the Continent the calculus was making unprecedented strides, crossing national boundaries. Among the first nationalities prominent in its development were the Swiss, as we have seen, then the French, and later the Germans.

Joseph Louis Lagrange (1736–1813), of French ancestry, was born and raised in Italy. By the time he was eighteen, he was teaching geometry at the Royal Artillery School in Turin, where he was born. Euler, while head of the Berlin Academy, received some notes on a "calculus of variations" that Lagrange had worked out. Euler had already written a book on this subject when Lagrange was still a boy. On recognizing the significance of these notes, Euler realized that here was a man of unusual mathematical ability. Not long thereafter, when Euler moved back to Saint Petersburg, he and d'Alembert saw to it that Lagrange, who was then only forty years of age, was appointed head of the Berlin Academy.

Lagrange's major research dealt with a systematization of the laws of mechanics that Galileo had pioneered. Lagrange was able to organize the whole field of theoretical mechanics by establishing very general equations. This was done entirely by analysis as opposed to geometry. The results were embodied in a famous treatise, the *Mécanique analytique* (1788), which, interestingly enough, although it dealt with mechanics, did not contain a single geometric diagram!

His interest in the laws of mechanics led Lagrange to investigate some unsolved problems of astronomy. Newton's theory of universal gravitation still had some weak spots: he had succeeded only with the problem of *two* bodies in motion simultaneously. But the solar system is comprised of *many* heavenly bodies. Lagrange succeeded in working out the mathematics that applied (in special cases only) to a system of three or more bodies simultaneously in motion, as, for example, the moon-earth-sun system.

On the death of Frederick the Great in 1786, Lagrange, finding his role in Berlin less than happy, accepted the invitation of Louis XVI to come to Paris and finish his mathematical work there as a member of the French Academy. Although he was received there with enthusiasm, he had come to feel that he could no longer contribute to mathematics. He had lost his enthusiasm and become melancholy. Forsaking mathematics, he turned to languages, to metaphysics, to medicine, chemistry, and religion. With the advent of the French Revolution, however, things changed for him. His zest for mathematics returned. He noted with keen interest the events of the Revolution but was horrified at the excesses of the Reign of Terror. A few years later he published two important books on the theory of functions. They were intended to make the calculus of Newton and Leibniz more "rigorous," that is, more logically perfect.

During the period of the Revolution, Lagrange was made head of a commission formed to create a new system of weights and measures. The result was the metric system, which was to become the universal tool of the world of science. In time, the system was adopted by virtually all nations for everyday uses as well. Now, almost two hundred years later, the United States is belatedly "going metric."

Figure 5.10 shows three French stamps honoring Lagrange, d'Alembert, and Laplace. The Russian stamp at the upper left commemorates the International Congress of Mathematics and shows the integral sign (∫) used in the calculus.

Fig. 5.10. Integral sign representing calculus—and three great Frenchmen who used calculus

Jean le Rond d'Alembert (1717–1783), a member of the Academy of Sciences, is noted chiefly for his contribution to gravitational theory; he sponsored the work of Lagrange and Laplace. He published an essay on hydrodynamics and wrote on complex numbers, calculus, differential equations, and Fourier series and was overly enthusiastic about analytical mechanics. He assisted in the preparation of Diderot's celebrated encyclopedia (1751–1772) in the early stages of that stupendous project.

Pierre Simon de Laplace (1749–1827), a contemporary of Lagrange, was a brilliant mathematical astronomer who contributed greatly toward the general acceptance of the new Newtonian explanation of the planetary motions, which even half a century after it had been published was still open to criticism. Through the help of the eminent mathematician d'Alembert, Laplace became professor of mathematics at the École Militaire in Paris while still scarcely twenty. His fame soon spread as he published in rapid succession a wide variety of research papers.

Whereas Lagrange was a superb mathematician, Laplace was primarily a physicist and an astronomer, for whom mathematics was a tool, albeit an extremely powerful one. His chief contributions were to astronomy. He spent a quarter of a century on research in astronomy, publishing between 1799 and 1825 five huge volumes under the title *Mécanique céleste.* This treatise was elaborate in scope, and its contents were highly abstract. Moreover, it was difficult reading, even for other scientists. Nevertheless, the treatment given by Laplace paved the way for greater insight into many areas, including gravitational theory, magnetic field theory, light, sound, hydrodynamics, heat conductivity, and electrostatics. Laplace also wrote a book for the general reader called *Exposition du système du monde,* which is a delightful popular explanation of astronomy. He also contributed importantly to the theory of probability. Here, as elsewhere, he frequently failed to acknowledge his indebtedness to other scientists, whose discoveries he simply incorporated with his own.

Living in the turbulent times of the French Revolution, Laplace was sufficiently shrewd to avoid coming to grief politically. He was a consummate opportunist: when the regime was Republican, he was an ardent Republican; when Napoleon came to power, he readily became an enthusiastic Royalist. He was more fortunate than his friend Lavoisier, the well-known chemist, who lost his head on the guillotine; Laplace kept his.

Here was a keen intellect whose personality was a strange mixture. A fair share of vanity, at times overly immodest, not always fair to his colleagues, he nevertheless was generous in encouraging younger, promising scientists.

Among other distinguished astronomers of the eighteenth and nineteenth centuries, we must also recall Sir William Herschel and his son John, as well as Francis Baily; all three are honored on the British stamp shown in figure 5.11, commemorating the 150th anniversary of the Royal Astronomical Society of England.

Although he was born in Germany, William Herschel spent most of his life in England, becoming one of her most renowned astronomers. He

Fig. 5.11. Francis Baily and William and John Herschel, pioneer members of the Royal Astronomical Society

made his own instruments and ground his own lenses, producing the finest telescopes of his time. A prolific observer, he is remembered for his discovery of a new planet, now known as Uranus; actually he rediscovered and described it completely. Herschel was the first astronomer to show that our solar system is but a tiny speck in the vast universe of stars. His son John also distinguished himself as an astronomer, systematically studying the stars in the Southern Hemisphere as his father had done in the Northern Hemisphere. John Herschel was one of the first scientists to apply photography to astronomy. Both Herschels were ably and conscientiously assisted in their work by Caroline Herschel, William's sister. The first woman astronomer of note, she discovered seven comets.

Francis Baily, one of the founders of the Royal Astronomical Society and many times its president, is remembered for his studies of the sun.

The century and a half during which modern mathematics emerged was fraught with dramatic advances in science and technology. A group of renowned scientists and mathematicians in England founded the Royal Society in 1662. Waterpower began to be used on a large scale about 1700, and the power loom had been invented a few years earlier; gas was first obtained from coal by distillation (1688); Fahrenheit devised the thermometer that goes by his name (1714); sulfuric acid began to be manufactured commercially in 1736, to the enhancement of many chemical industries. By the middle of the eighteenth century the Industrial Revolution was in full swing: iron wheels were being used on coal cars (1755); the manufacture of cement had begun (ca. 1756); Watt's steam engine came into general use as a prime mover (1785); Cartwright gave us the steam-power loom (1785); ships were being equipped with screw propellers (1785); and Whitney's cotton gin (1793) was to affect the course of American history decades after its invention.

Modern mathematics and science had arrived: the world was on the brink of new horizons.

New Horizons
1800–1900

As the world moved into the nineteenth century, science and technology were making rapid advances. Precision measuring instruments were developed, the slide rule was improved, and the world of science adopted the metric system. The era of electricity was soon to open, and as the American Civil War began, Charles Darwin's epoch-making theory of evolution shocked people on both sides of the Atlantic.

In mathematics, the dawn of the nineteenth century had already been heralded by the immortal Carl Friedrich Gauss, whose genius pervaded all branches of mathematics. Outstanding mathematicians zealously expanded the paths that had been blazed during the previous century and a half. New ideas proliferated. The creation of non-Euclidean geometry was to change the character of mathematics drastically.

THE LIBERATION OF MATHEMATICS

The revolutionary change in mathematics, which took place near the middle of the nineteenth century, was anticipated by Carl Friedrich Gauss (1777–1855), often referred to as the Prince of Mathematicians. Born in Brunswick, Germany, a year after the signing of the Declaration of Independence in America, he was soon recognized as a child prodigy.

Befriended by the Duke of Brunswick, Gauss at the age of fifteen entered a private school, where he soon mastered by himself the works of Newton, Euler, and Lagrange. At the age of eighteen he entered the University of

Göttingen, where he became friendly with Wolfgang Farkas Bolyai, scion of an old Hungarian family and later a famous mathematician in his own right. By the time Gauss was twenty-two, he had embarked on the field of his greatest talent, the theory of numbers. Shortly thereafter, in 1801, he published his brilliant research, the famous *Disquisitiones Arithmeticae*, an exhaustive treatise on the theory of numbers. On the basis of this book alone he is entitled to immortality. Yet, strangely enough, when he had sent the work to the French Academy the year before, it was rejected; Gauss had been far in advance of the judges. It was years later before the mathematical world realized that he had revolutionized the entire theory of numbers.

At the age of twenty-four Gauss reached a turning point in his life. By a strange quirk of fate, he became intensely interested in astronomy. The unexpected discovery of the asteroid Ceres (thought at the time to be a new planet) presented a problem: What was the form of its orbit? Only a few observations were available, but Gauss tackled the problem. He devised new procedures, including the method of least squares, and succeeded in predicting the orbit of Ceres with remarkable accuracy. Later the "planet" was found just where it was expected to be according to his calculations. With this accomplishment, his fame as an astronomer spread beyond his homeland.

Thus by accident Gauss was diverted from pure mathematics for a good part of his career. Not only did he become engrossed in astronomy, he also plunged into a number of other fields. In mathematical physics he dealt with problems of optics; he concerned himself with the design of astronomical instruments; he became involved in extensive geodetic surveys, making good use of the method of least squares, which he had devised as a young man. Stimulated by the needs of mapmaking, he contributed to the mathematical theory of surfaces; he made original studies of statistics and probability; he made many studies of terrestrial magnetism. In 1833, together with his friend, the physicist Wilhelm Weber, Gauss built a two-wire telegraph line of copper, which stretched over housetops for 2.3 kilometers. The line was used by the two friends for several years, but they never did pursue their invention further.

Despite his activities in these diverse fields, Gauss never completely forsook pure mathematics. Among his many bold contributions were his ideas about non-Euclidean geometry, about which we shall say more shortly. Equally as comfortable with pure mathematics as with applied mathematics, Gauss had miraculous powers of calculation and unusual powers of concentration. He had a talent for languages, taking up the study of Russian by himself at the age of sixty-two. He was reluctant to publish his discoveries, often keeping his notes in a drawer and not disclosing them to anyone. He died quietly of a heart attack in 1855 at the age of seventy-eight. On the centennial of his death, the German Federal Republic issued a commemorative stamp (fig. 6.1), the first honoring this, one of the greatest mathematicians of all time. At this writing, two more stamps honoring Gauss have appeared: one from the German Democratic Republic depicting Gauss and symbols of geodesy, the other from the German Federal Republic depicting the plane of complex numbers. (Complex numbers are also called Gaussian integers.)

Fig. 6.1. Gauss and Abel

In passing, we must not overlook the all-too-brief career of Niels Henrik Abel (1802–1829), who is shown in the Norwegian stamp in figure 6.1. Like the French mathematician Evariste Galois (who was killed in a duel at the age of twenty-one), the Norwegian mathematician Abel died tragically young. At the age of twenty-six his potentially brilliant career was cut short by tuberculosis. He showed little talent for mathematics until he was sixteen, but in those ten remaining years he became an apostle of the spirit of modern mathematics. In brief, where Newton and Leibniz had developed the calculus and men like Euler, Lagrange, and Laplace had put it to such good use, it remained for Abel to begin laying a rigorous, logical foundation under mathematical analysis. In addition, he contributed many new concepts, particularly on infinite series and elliptic functions. Perhaps even more significant are his famous "impossibility proofs," one of which demonstrates that whereas general algebraic formulas exist for the solution of equations up to and including the fourth degree, no general algebraic formulas can exist for equations of the fifth degree or higher. The tragedy of Abel's life was compounded by the bitter competition that prevailed among pioneers in mathematical research in the early nineteenth century. Partly because of this competitive climate in the mathematical community, Abel unfortunately never received recognition in academic circles, despite his many contacts with leading mathematicians of his day.

We come now to one of those rare, momentous turning points in the history of ideas: the emergence of non-Euclidean geometry. Anticipated by Gauss, this development, independently arrived at almost simultaneously by Nikolai Ivanovich Lobachevski and János Bolyai, marks the beginning of the axiomatic viewpoint in contemporary mathematics. Let us see what this means.

The system of Euclidean geometry inherited from the Greeks was regarded for nearly two thousand years as a "perfect" logical system. Yet there was one serious weakness—the so-called parallel postulate. This axiom stated that through a given point P outside a given line l, one and only one line (MN) can be drawn parallel to the line l (fig. 6.2).

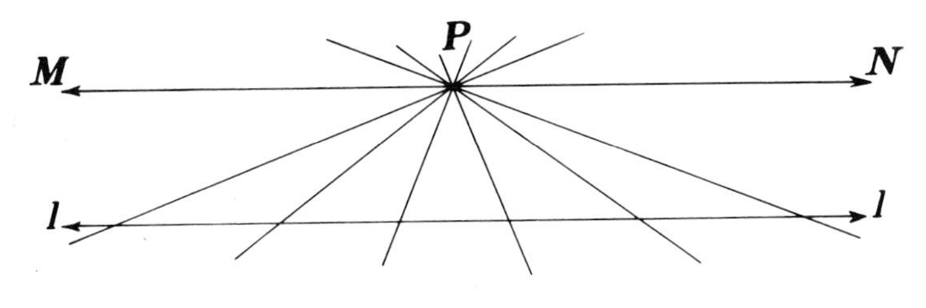

Fig. 6.2

Unlike the other axioms, this one was not really "self-evident," for it involved the concept of the parallelism of lines of infinite extension. Over the years, many mathematicians had attempted to prove it by referring to the other, simpler axioms, but to no avail.

As early as 1792 Gauss began to think about this puzzling question, discussing it from time to time with his friend Wolfgang Farkas Bolyai. Gauss made notes about his own ideas and put them aside for some future time. He already suspected that the parallel axiom could not be proved and, further, that it was possible to construct a logically consistent geometry by using a different parallel axiom. Wolfgang Bolyai returned to Hungary thinking he had found a proof, but when subsequent correspondence with Gauss revealed that the proof was unsatisfactory, Bolyai gave it up as a complete failure, even discouraging his son János from pursuing the vexing problem further.

But János persisted in his efforts. At the age of twenty-one the brilliant young mathematician had grasped the elusive concept: the Euclidean parallel axiom was *not* necessary to a system of geometry. He created what he called the *absolute science of space* by simply ignoring the traditional parallel axiom. By 1823 his ideas had crystallized, and he wrote joyously to his father: "I have created a new universe out of nothingness." At his father's urgent insistence, János published his twenty-six page essay as an appendix to a book his father had written. The young Bolyai had independently created a new geometry with a new parallel axiom. Ironically, Gauss had attained essentially the same result some twenty or thirty years earlier but had lacked the courage to publish such revolutionary ideas.

While this was happening in Hungary and Germany, a professor of mathematics at the University of Kazan in Russia, Nikolai Lobachevski, had also been pondering the same question for a number of years. By 1826, he was convinced that a new geometry could be built without Euclid's parallel axiom. Such a geometry would not be quite the same as Euclidean geometry, but it would nevertheless be a geometry that was logically self-consistent. He called it *pangeometry*. Instead of only *one* line through point *P*, he assumed that at least *two* lines could be drawn parallel to line *l*. Keeping the other axioms of Euclid, he built his new geometry. To be sure, like that of Bolyai, this new geometry was indeed strange. It seemingly defied common sense or intuitive experience. For example, in it the sum of the angles of a triangle is *less* than 180°; there are infinitely many parallels through a given point; and if two lines are both perpendicular to a third line, they are not parallel, nor do they intersect. No wonder it took courage to publish such daring, unorthodox ideas! His results were published between 1826 and 1829, preceding the publication of Bolyai's essay. Hence this system, known as hyperbolic geometry, is generally credited to Lobachevski. Neither was aware of what the other was doing; both arrived at the same idea.

The Hungarian stamp in figure 6.3 (*center*) portrays Wolfgang Farkas Bolyai, and the Romanian stamp on the left shows his son, János Bolyai. The Russian stamp pictures Nikolai Lobachevski.

Fig. 6.3. János Bolyai, Farkas Bolyai, and Nikolai Lobachevski

That three men—Gauss, Lobachevski, and Bolyai—should have arrived independently at the same undreamed-of innovation is in itself remarkable enough. But the real significance is the profound effect it had on the subsequent development of mathematics, especially in view of Georg Riemann's creation twenty years later of another type of non-Euclidean geometry in which there are no parallel lines at all!

The acceptance of these non-Euclidean geometries changed the very foundations on which mathematics rested. Accordingly, any system of mathematics is what mathematicians wish to make it. They select the axioms and build on those axioms. The resulting systems may not be compatible with each other, but each is logically consistent within itself. In short, mathematics could no longer be regarded as a body of truths inherent in the design of the universe: instead, mathematics is a human creation. This breakthrough—this liberation—was to the evolution of mathematics what the Copernican dictum "Sun, stand thou still!" was to the history of astronomy—a truly revolutionary change.

But there is more to the story of the liberation of mathematics, for a similar axiomatization of algebra was also taking place during the middle of the nineteenth century. It was precipitated by an eccentric Irish mathematician, Sir William Rowan Hamilton (1805–1865). A precocious child, he was advanced in arithmetic at the age of three; he could read Latin, Greek, and Hebrew at the age of five and understood French and Italian at eight. Unfortunately, he was brought up by a doting uncle, an Orientalist, at whose insistence he misspent the next five years on Oriental languages, including Persian, Arabic, and Sanskrit.

After several false starts, culminating in an unsuccessful attempt to become a professional astronomer, Hamilton found his métier in mathematics. Today he is remembered principally for (1) his contributions to the theory of optics, where he expounded his "systems of rays" (which, curiously enough, fit neatly into the wave mechanics associated with modern quantum theory and the theory of atomic structure) and (2) his contributions to the theory of algebra, in which he created his so-called quaternions. Briefly, Hamilton developed a new kind of algebra in which the commutative law of multiplication no longer holds—that is, $a \times b$ is not equal to $b \times a$; instead, $a \times b = -b \times a$. This denial of the commutative property was a courageous breakthrough, or in his own words, a "monstrous innovation." He wrestled with this problem for more than twenty years, extending the notion of a complex number $a + bi$ as a *couple* to the idea of a *quaternion*—that is, a number with four parts, $a + bi + cj + dk$. He was quite carried away with the importance of his invention. Recalling the discovery years later, he wrote to his son:

> On the 16th [October 1843] . . . as I was walking along the Royal

> Canal with your mother . . . an undercurrent of thought was going on in my mind . . . which gave at last a result, whereof it is not too much to say that I felt at once the importance. An electric circuit seemed to close, and a spark flashed forth. . . . Nor could I resist the impulse—unphilosophical as it may have been—to cut with a knife on a stone of Brougham Bridge, as we passed it, the fundamental formulae $i^2 = j^2 = k^2 = ijk = -1$.

He appears, appropriately enough, on companion Irish stamps, one of which appears in figure 6.4.

Fig. 6.4. William Rowan Hamilton

Essentially, Hamilton showed that it was possible to construct a logically consistent algebra that defied the commutative law of multiplication. This alone was a breakthrough nearly as momentous as the creation of non-Euclidean geometries. Today quaternions are no longer of importance, having been supplanted first by vector analysis and then by the more general tensor analysis, which was developed in the early twentieth century. Nevertheless, Hamilton blazed the path that led to many different algebras, just as there are many different geometries.

We also take note of several other eminent mathematicians and scientists of the latter part of the nineteenth century: Sonya Kovalevski (1850–1891), Pafnuti Chebyshëv (1821–1894), Lambert Adolphe Quételet (1796–1874), and Jean Foucault (1819–1868).

The name of Sonya Kovalevski is always associated with her friend and teacher, the great German mathematician Karl Weierstrass, who contributed much to the theory of functions. Born in Moscow, Kovalevski began studying mathematics at fifteen, went to Berlin at twenty to continue her studies under Weierstrass, and took her doctorate at the age of twenty-four. A strikingly beautiful woman, she was torn between her devotion to mathematics and a desire for a normal social life. This conflict, together with an ill-fated marriage, all but wrecked her professional career. Eventually, however, her contributions to analysis and function theory won her the international recognition she deserved. Two years before she died at the age of forty-one, she became professor at the University of Stockholm, where she had been lecturing for the previous five years.

Chebyshëv (also Tchebycheff) was professor at the University of Saint Petersburg. Although he was younger than Lobachevski by nearly thirty years, they were generally regarded as rivals. Chebyshëv's contributions range over a wide field: probability, theory of integration, linkages, and especially the theory of numbers. In this latter field, for example, he was able to prove that if $n > 3$, there always exists at least one prime number between n and $2(n - 1)$.

Sonya Kovalevski, whose portrait appears on one of the Russian stamps in figure 6.5, was one of the few famous women mathematicians of modern times. Chebyshëv appears on companion Russian stamps, one of which appears in figure 6.5. Quételet is shown on the Belgian stamp in figure 6.6, and the distinguished physicist Foucault is shown on the accompanying French stamp.

Fig. 6.5. Sonya Kovalevski and P. L. Chebyshëv

The Belgian mathematician Quételet, for nearly half a century director of the Royal Observatory in Brussels, was noted both as an astronomer and as a statistician. His greatest contribution was in the application of mathematical methods to the study of vital statistics. In this connection he explored the bell-shaped probability curve and pioneered in the mathematical basis of insurance theory.

The French physicist Jean Foucault (1819–1868) distinguished himself chiefly by his measurements of the velocity of light and by his famous pendulum experiment, which vividly demonstrated the earth's rotation on its axis. Earlier attempts to measure the velocity of light had been made by the Danish astronomer Olaus Roemer (ca. 1675) and by the English astronomer James Bradley (ca. 1725); both men used astronomical methods. In the 1850s Foucault, improving on a method of Armand

Fizeau's, succeeded by physical measurements in determining the velocity of light to be approximately the same at that later obtained in 1882 by Albert Michelson, namely, 299 853 kilometers a second.

Fig. 6.6. Lambert Adolphe Quételet and Jean Foucault

Making use of the fact that a swinging pendulum tends to remain in the plane of its oscillation, Foucault suspended a heavy iron ball about sixty centimeters in diameter from the dome of a church by a steel wire about sixty meters long. As the great pendulum continued swinging, a spike at the bottom of the ball made marks in the sand on the floor. These marks changed direction as time went on, showing that the earth rotated beneath the pendulum. It was the first direct physical evidence that the earth rotated on its axis. A so-called Foucault pendulum can be seen in the entrance of the United Nations building in New York City and in the Smithsonian Institution in Washington, D.C. Foucault also arrived at the theory of the gyroscope.

MEASUREMENT AND COMPUTATION

Civilization as we know it today is completely dependent, among many other things, on the art of measurement. Even in early times people recognized the importance of measuring the distance between two places, the areas of fields, the capacity of containers, the weights of coins, and so on. These needs called for relatively crude measures. A need for greater precision arose in navigation and astronomy, where the first advances in devising more precise instruments were made.

But it was the rise of experimental science and the advent of the Industrial Revolution that made precision measurements indispensable. Contemporary science and technology simply could not have reached their high levels of attainment without the use of precision instruments. One of the primary needs of industry is the measuring of machine tools and machine parts.

Measuring instruments used by technicians in shop work are shown on many stamps: the U.S. stamp at the top of figure 6.7 shows a machinist handling a micrometer caliper; the Indian stamp to the right shows an "outside caliper" used to measure the diameter of a cylinder or a sphere; the lower stamp, from the Republic of Korea, shows a slide micrometer.

Fig. 6.7. Shop measuring instruments *(clockwise from top)*: micrometer caliper, outside caliper, slide micrometer

Precise measuring is by no means limited to industry and machinery; land measurement is as important today as it ever was, if not more so. As we have seen in an earlier chapter, the art of surveying originated in ancient Egypt. Later the Romans improved on the surveying instruments they had acquired from the Egyptians. Telescopic sights, spirit levels, micrometer microscopes, and vernier calipers—all of which permitted more precise measurements—were developed in the seventeenth century in response to the needs of astronomy. By the late eighteenth century, these features had been incorporated in the theodolite and the transit, modern surveying instruments by means of which horizontal as well as vertical angles are measured with great precision. The theodolite is the basic instrument; in general, it is more precise than the transit. The theodolite is pictured on the Finnish stamp (*lower right*) in figure 6.8, and a surveyor using a transit level can be seen on the German stamp above it.

Fig. 6.8. Surveying (*clockwise from left*): aerial mapping, transit level, theodolite

In modern times many refinements have been made in surveying techniques—as for example, mapping from aerial photographs, as suggested on the Finnish stamp to the left in figure 6.7. More recently, electronic devices, including the laser, are being used for measuring distances; satellites and electronic computers assist with geodetic surveys.

Surveyors measure direction, horizontal and vertical angles, and horizontal and vertical distances. For vertical heights or elevation, they use leveling and a gauge, as shown on the Nigerian stamp in figure 6.9. For increasing their accuracy in measuring horizontal distances and angles, they use triangulation methods, involving a series of connected triangles or quadrangles, as suggested on the East African stamp below it.

When making land measurements within an area less than twelve miles square, surveyors need not take the earth's curvature into account. This is called plane surveying, and it applies to measuring real estate, small farms, highway construction, and the like. Larger portions of the earth's surface cannot be regarded as a plane, and more elaborate procedures must be used. This is known as geodetic surveying; it is used for determin-

Fig. 6.9. Surveying: leveling and triangulation

ing political boundaries, for accurate mapping of rivers and coastlines, and so on.

Almost up until the beginning of the nineteenth century the weights and measures in common use in Europe comprised a hodgepodge of unrelated systems. Shortly after the French Revolution, the newly created metric system was recognized officially, although it did not become mandatory in France until 1840. The scientific community soon adopted it universally, and as time went on, one nation after another adopted it for general and commercial uses as well. The United States was one of the few nations that lagged behind in this respect.

In 1960, delegates from thirty-six nations, including the United States, agreed on a new physical standard for the international meter: 1 meter = 1 650 763.73 orange-red wavelengths of an atom of krypton-86 gas in a vacuum. This is the basis of the so-called SI system, or *Système International d'Unités*. In this system the fundamental units are the following:

Length	*meter*	m
Mass	*kilogram*	kg
Time	*second*	s
Electric current	*ampere*	A
Temperature	*kelvin*	K
Luminous intensity	*candela*	cd

The symbols for these units are exhibited by the Romanian stamp (fig. 6.10, *upper left*), where the metric system was introduced in 1866. The Korean stamp below it is reminiscent of the earlier metric units and shows a cross section of the standard metal meter bar in the upper right-hand corner. On the East African stamp (*upper right,* one of a set of four stamps), the relation between pound and kilogram is shown. Yugoslavia also publicizes the metric system (*lower right*).

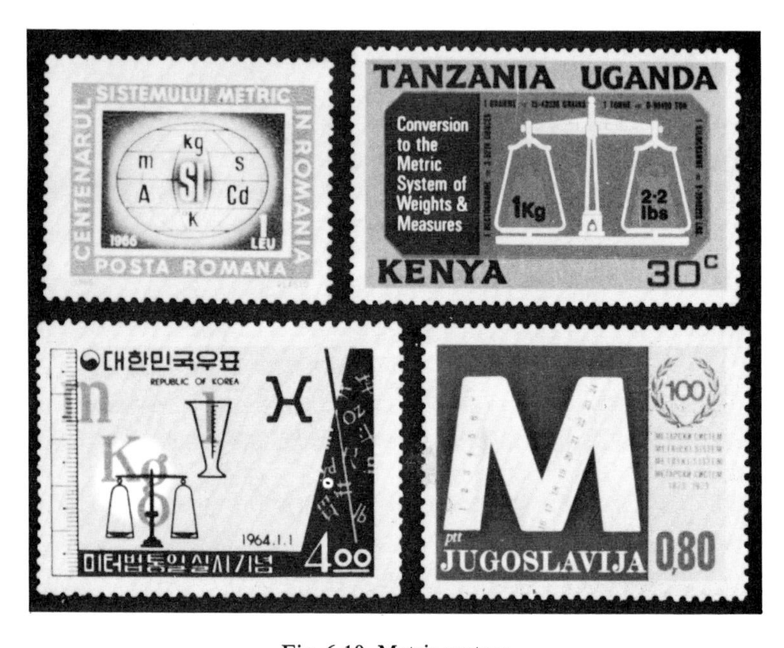

Fig. 6.10. Metric system

By an act of Congress in 1866, the use of the metric system was made legal in the United States. Now, after more than a hundred years, we are moving in the direction of mandatory metric measures. The Metrication Act of 1972 instructs the president to appoint a metrication board to plan and supervise a ten-year conversion program. Presumably this legislation will, in effect, cause compulsory metrication, and so the United States will eventually abandon the foot, pound, and quart. Stamps of several nations have recognized metric conversion in other parts of the world. We show three of them (fig. 6.11): the center one from Pakistan and the others from Australia, two of a set of four whose cartoon style has provoked both criticism and chuckles.

Since the earliest times, people have always made use of some kind of mechanical device to help them in computing, from the Roman pebble board to the abacus and the counting table. About 1630, William Oughtred

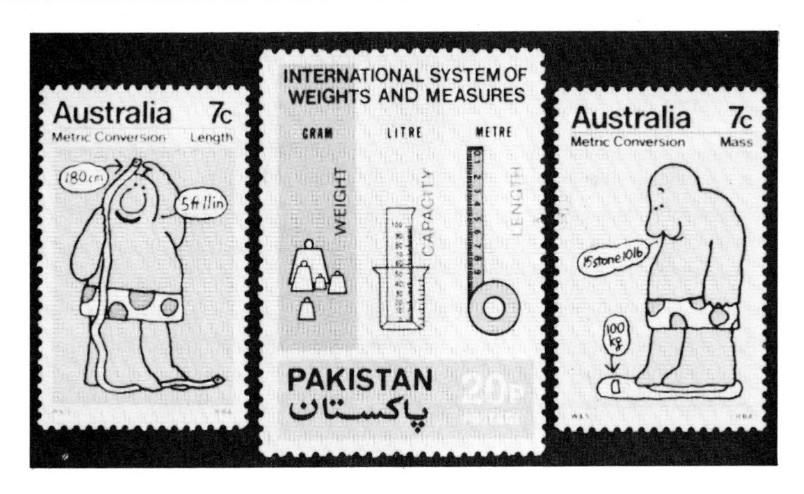

Fig. 6.11. Metric units

and Edmund Gunter, two English mathematical instrument makers, adapted Napier's logarithms (1614) to appropriate scales on a circular rule to assist their computations. Some two hundred years later the French army officer Mannheim gave the slide rule its modern form, as suggested on the Romanian stamp in figure 6.12 (there is also a companion stamp). The Colombian stamp to the right shows, among other computation symbols, the multiplication sign, ×, popularized by Oughtred.

Fig. 6.12. Computation: slide rule; computation symbols

Among the earliest mechanical calculators you may recall the machine built by Pascal (1642), with which one could do problems in addition by simply turning a handle. This adding machine was a model for our modern improved Comptometers. Whenever you look at the cyclometer on your bicycle, the odometer on your car, or the automatic counter on a mimeograph machine, you should be reminded of Pascal's adding machine.

This early calculator was followed by Leibniz's calculator (1671), which could multiply and divide as well as add and subtract. One hundred and fifty years later, Charles Babbage tried to build a "difference engine" to be used mostly for navigational and astronomical computations. From 1820 to 1856 he struggled to complete such a machine, but with no success. Curiously enough, in the 1920s scholarly research revealed that as early as 1624 one Wilhelm Schikard, a German mathematician and astronomer, had communicated to Kepler plans for a calculating machine. When reconstructed in 1960, the machine was found to be functional. It is shown in figure 6.13 on the German Federal Republic stamp to the right. The Danish stamp shows a conventional office calculator and ledger.

Fig. 6.13. Early calculating machines

THE AGE OF ELECTRICITY

When in 1752 Benjamin Franklin conducted his experiment with a kite during a thunderstorm and proved that lightning was nothing but a discharge of static electricity, he opened up an entirely new world. Scientists, experimenting with friction machines and Leyden jars during the latter half of the eighteenth century, had regarded electricity as a fluid of some kind.

Some thirty years later, Charles Augustin de Coulomb (1736–1806), a French physicist, invented the torsion balance, which could measure the quantity of a force by the amount of twist it produced in a thin, stiff fiber. With this delicate instrument Coulomb was able to show that the attractive force between two electrically charged spheres varied directly as the product of their charges and inversely as the square of the distance between them. This is Coulomb's law, and it is analogous to Newton's formula for the gravitational force between two bodies:

$$F = \frac{m_1 m_2}{d^2}$$

Franklin's immortal experiment is commemorated on a U.S. stamp (fig. 6.14, *left*); issued on the 250th anniversary of his birth, it was entitled "Taking Electricity from the Sky." The French stamp (*center*) shows Coulomb and his torsion balance, and the stamp of Italian Somaliland (*right*) portrays the Italian physicist Alessandro Volta, commemorating the centennial of his death.

Fig. 6.14. Franklin; Coulomb; Volta

The pioneering efforts of Franklin and Coulomb were concerned only with *static* electricity. The two men could scarcely have foreseen that during the nineteenth century the study of electricity was to serve as a great stimulus to mathematical investigations. It was not until about 1800 that Alessandro Volta (1745–1827) invented the *voltaic* pile, or battery, which made him famous. This was the first time that current electricity was produced by other than natural means. Volta was encouraged in his work by Coulomb; he received much acclaim, and in his honor, the unit of electromotive force is named the *volt*.

The next significant advance was made by a Danish professor of physics and chemistry, Hans Christian Oersted (1777–1851), who discovered that when a compass needle was placed near a wire through which an electric current was passing, the compass needle was deflected, pointing at right angles to the wire. Oersted had thus demonstrated that magnetism and electricity were related, and the study of electromagnetism was born. Excitement was rampant, and many scientists became involved: Ampère, Arago, Michael Faraday, and Joseph Henry, among others.

André Ampère (1775–1836), a French physicist, chemist, and mathematician, was the first to apply advanced mathematics to electrical and magnetic phenomena. He enunciated the concept of lines of magnetic force in the neighborhood of a conductor carrying a current. He was also the first to point out the difference between the *amount* of current flowing through a conductor and the *force* driving the current along; the former is measured in amperes; the latter, in volts. Ampère also contributed to the theory of advanced calculus.

A contemporary of Ampère's, the French physicist François Arago, began to explore the properties of electricity after studying the properties of light. It was he who showed that a copper wire carrying a current was a true magnet and that iron was not essential to creating magnetic force.

A portrait of Oersted appears on the Danish stamp (fig. 6.15, *left*), and the 150th anniversary of Oersted's discovery of electromagnetism is commemorated on the Danish stamp to the right. Ampère, together with Arago, appears on the French stamp in the center.

Fig. 6.15. Oersted; Ampère and Arago; electromagnet

We digress for a moment to refer to two scientists whose prime concern was not with electricity. Gustav Kirchhoff, a German physicist who proved that the speed of an electrical impulse was equal to that of light, is best remembered for his work in photochemistry and spectroscopy. By means of the spectroscope he was able to identify many of the chemical elements in the sun. Perhaps Kirchhoff's greatest claim to fame was his discovery (ca. 1870) of "blackbody radiation." An ideal blackbody absorbs all the radiation falling on it and reflects none. For experimental purposes, the blackbody is an almost completely enclosed cavity in an elongated, hollow metal cylinder. This concept of blackbody radiation was to lead to Planck's quantum theory thirty years later.

The Austrian physicist Ludwig Boltzmann contributed (ca. 1880) significantly to advances in two areas of science. In connection with the kinetic theory of gases, he used advanced mathematics and statistical theory to explain the behavior of gases and some of the principles of thermodynamics. The kinetic theory explains the phenomena of heat and pressure in terms of the continuous haphazard motion and elastic collisions of atoms and molecules in liquids and gases. Furthermore, using sophisticated mathematics in connection with radiation theory, Boltzmann enunciated the so-called Stefan-Boltzmann law, which states that radiation varies directly with the fourth power of the absolute temperature. He based his calculations on the experimental results of Joseph Stefan.

An excellent portrait of Kirchhoff appears on the German stamp (fig. 6.16) commemorating the sesquicentennial of his birth. The Nicaraguan stamp to the left diagrammatically depicts a four-cylinder gasoline engine in honor of Boltzmann's contributions to thermodynamics.

Fig. 6.16. Boltzmann's law; Kirchhoff

Returning to the progress made in electrical theory as the century drew to a close, we note that two names stand out: Maxwell and Hertz. Scottish-born James Clerk Maxwell (1831–1879) became professor of mathematics at Aberdeen University in 1856. Early in his career he was recognized for his work in astronomy, particularly in connection with Saturn's rings. Shortly thereafter he became professor of experimental physics at Cambridge University, and it was about this time (1871) that he carried on his brilliant work in electromagnetic theory. By means of a few relatively simple equations, Maxwell was able to express the properties of a magnetic field, an electric current, and an electric field; indeed, he showed that magnetism and electricity were intrinsically interrelated.

Maxwell's equation, together with a representation of wave propagation, are shown on the Nicaraguan stamp in figure 6.17.

Fig. 6.17. Maxwell's law

Maxwell also showed conclusively that there existed a wide range of electromagnetic radiation wavelengths, of which visible light constituted only a small part, with ultraviolet and infrared wavelengths on either side of the visible spectrum. Although his concept of a space-filling luminiferous ether as a medium for transmitting electromagnetic waves was doomed to oblivion by the work of later scientists, his fundamental equations were found to be valid even after Einstein's relativity theory had upset most of classical physics.

Heinrich Hertz (1857–1894) was a German physicist who in the 1880s became deeply interested in electromagnetic theory and specifically in the equations worked out earlier by Maxwell. By 1890 Hertz had discovered electromagnetic waves of long wavelengths, which he produced by using sparks. These so-called Hertzian waves formed the basis of Marconi's "wireless" communication and are the basic waves of radio transmission. Toward the end of his career Hertz worked with cathode rays, but he did not live long enough to see the birth of modern radio broadcasting or the discovery of the electron. He died prematurely at the age of thirty-six.

Hertz is pictured (fig. 6.18) on stamps of the German Democratic Republic, German Federal Republic, and Czechoslovakia. The latter also depicts an electromagnetic field.

The nineteenth century, in a sense, was the age of steel. The railroads and the steamboat appeared on the scene; so, too, did steam and water

Fig. 6.18. Heinrich Hertz

turbines and the internal combustion engine. Electric power had arrived: the dynamo, the motor, the electric lamp, the telephone, and the telegraph. Photography and printing came of age.

Specificially, some of the highlights that stand out include Fulton's side-paddle steamboat (1803), Braille's printing for the blind (1829), daguerreotype photography (1839), the first steel-cable suspension bridge (1840), Howe's high-speed sewing machine (1845), commercial production of aluminum (1855), the Bessemer converter for making steel (1856), ammonia refrigeration (1860), the Westinghouse automatic air brake (1872), the Otto four-cycle gasoline engine (1876), the first steel-frame skyscraper (Chicago 1884), the Mergenthaler linotype printing machine (1884), the Parson's steam turbine (1884), Eastman's hand camera (1886), and Edison's motion-picture projector (1895).

The industrial revolution in America was full blown.

The Dawn of a New Century

1900–1940

THE twentieth century opened uneventfully. There was little to foreshadow the unrest and ferment that was to come in the next few decades—World War I, the Russian Revolution, the Great Depression. Science and technology were making great strides. It was the era of Freud, Jung, and Adler. Vitamins made their debut in 1912, Fleming gave us penicillin in 1929, and sulfa drugs first appeared in 1932. High-speed tool steel became available in 1900; nitrogen fixation, 1903–8; synthetic resins (Baekeland), 1906; the Owens automatic bottle-blowing machine, 1904; the Model T Ford, 1909; the Sperry gyroscope, 1910; sound movies, 1927; and jet-propelled aircraft, 1930.

In mathematics, the leaven that had been provided by non-Euclidean geometry, by the abstract algebra of Hamilton, and by George Boole's symbolic logic was working. Mathematicians, led by Peano, Frege, Cantor, Veblen, Hilbert, and others, were bent on a relentless quest for the *structure* of mathematics. This meant more attention to axiomatic systems, with increased abstraction and logical rigor.

PURE AND APPLIED MATHEMATICS

Henri Poincaré (1854–1912), one of the foremost mathematicians at the turn of the century, was a brilliant analyst, a mathematical physicist, and an able interpreter of the philosophy of science. In addition, he was one of the very few mathematicians who have attempted to explain psychologically how the mind of a mathematician works. According to Poincaré, creative thinking may very well go on subconsciously between periods of conscious activity; the conclusion or solution to a problem may come quite unexpectedly at some later time.

Poincaré was the last of the great universalists, equally at home in almost all branches of mathematics. Today no mathematician can be familiar with every area, for the field is so vast and has been so deeply cultivated that specialization is unavoidable.

One of the most remarkable mathematical minds of modern times was that of Srinivasa Ramanujan (1887–1920). Born of humble parents in India and with only a simple schooling available, he entered the University of Madras at the age of sixteen, teaching himself all the mathematics that he could without taking the conventional university courses. In due time his unusual talent for mathematics was recognized, and through the help of friends he went to England as a protégé of the distinguished British mathematician and number theorist G. H. Hardy.

The fact that Ramanujan's formal mathematical training left much to be desired only served to highlight his unusual gift of extraordinary insight into the properties of numbers, continued fractions, and partitions. Someone has said of him that every positive integer was one of his personal friends. Once, while he was ill in bed, Hardy paid him a visit, remarking that he had ridden in a taxicab whose number was 1729 and that the number seemed rather a dull one.

"No," was Ramanujan's instant reply. "It is a very interesting number. It is the smallest number expressible as the sum of two cubes in two different ways."

$$1^3 + 12^3 = 1729 = 9^3 + 10^3$$

In 1917 he became ill and never recovered. He returned to India two years later and died shortly thereafter at the age of thirty-three.

In their search for structure and logical consistency, early twentieth-century mathematicians were plagued (among other things) by the concept of infinite classes. At the height of these activities, Alfred North Whitehead and Bertrand Russell, both brilliant philosophers and mathematicians, published their monumental three-volume work, *Principia Mathematica* (1910–13). This was an ambitious attempt to place the entire field of mathematics on an axiomatic basis; in short, they regarded all mathematics simply as a branch of logic. Bertrand Russell, a profound logician and philosopher as well as a consummate mathematician, was widely known for his liberal and unconventional views on politics, religion, education, sex, and social problems. He left a permanent influence on the philosophy of mathematics.

Poincaré is portrayed on the French stamp to the right in figure 7.1; Ramanujan appears on the lower Indian stamp, and Bertrand Russell's portrait is given on the upper one.

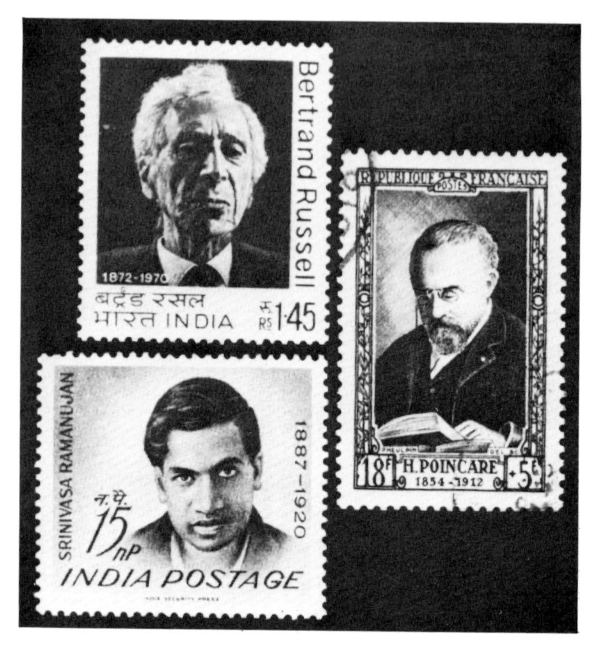

Fig. 7.1. Russell, Poincaré, and Ramanujan

The work of several other scientists honored on stamps, although not strictly "applied mathematics," is noteworthy. The French mathematical physicist Jean Perrin, professor of physical chemistry at the University of Paris for thirty years, is known for his work on Brownian motion. He was able to show how the approximate size of molecules and atoms could be calculated, thus giving real existence to heretofore fictional entities. Another French physicist, Paul Langevin, noted for his work in the electronic theory of magnetism, was also responsible for developing the principle of sonar, the ultrasonic sound-wave device for detecting submarines, the contours of ocean bottoms, and so on. The Hungarian physicist Baron Loránd Eötvös is known for his improved torsion balance to measure gravity as well as for his work in physical chemistry, geodesy, and geology.

These three scientists have been honored on stamps (fig. 7.2): Perrin (France, *right*), Langevin (France, *left*), and Eötvös (Hungary, *center*). The latter stamp also depicts his torsion balance.

The years at the turn of the century teemed with scientific developments.

Fig. 7.2. Langevin, Eötvös, and Perrin

Wilhelm Roentgen discovered the X ray about 1896, and the medical profession quickly recognized its value. Michael Pupin, the Yugoslavian physicist who came to America in 1874 (about the time that Alexander Graham Bell invented the telephone), would improve long-distance telephony by the end of the century by means of loading coils.

Roentgen appears on a stamp of the German Federal Republic (fig. 7.3, *upper left*) and one of Spain (*upper right*). Alexander Graham Bell's portrait is found on a stamp of the Republic of Niger (*lower left*), and Pupin

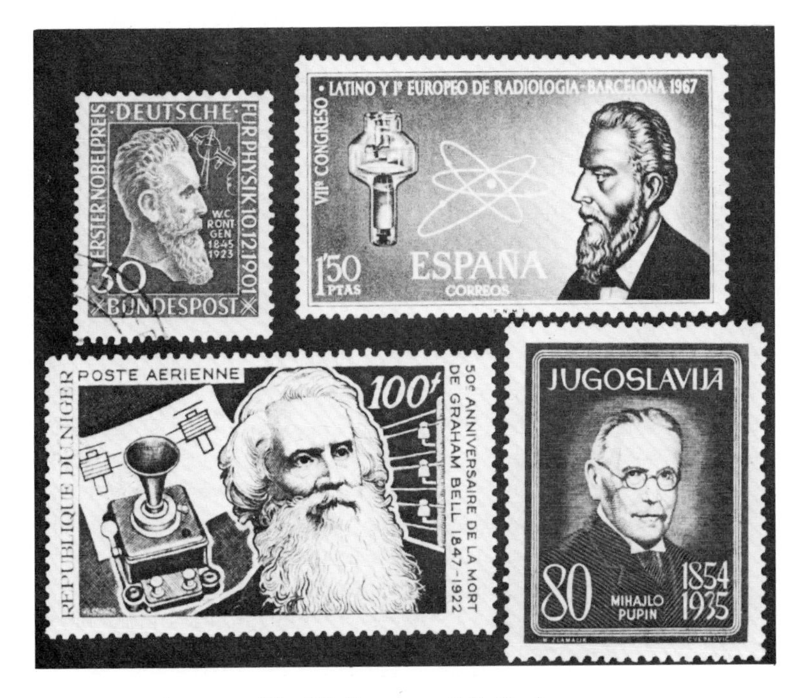

Fig. 7.3. Roentgen; Bell; Pupin

appears on one of Yugoslavia. Indeed, in 1976, the centennial of the invention of the telephone, literally hundreds of stamps were issued honoring Bell.

Nikola Tesla, an eccentric Hungarian inventor who came to this country in 1884, worked with Edison for a short period, but then struck out on his own. Through his efforts, electrical power really became practical; the transformer he designed, which required alternating rather than direct current, could raise the current to higher voltages for more efficient transmission over long distances. This caused bitter quarrels between Edison and Tesla. Edison would have nothing to do with alternating current.

Tesla appears on two stamps in figure 7.4. The other three stamps, all Yugoslavian, depict (*from the top*) an induction motor, a transformer, and an electronic control device.

One of the most brilliant scientists of this era was the British physicist Ernest Rutherford (1871–1937), who early in his career became interested in the newly discovered radioactivity, which had been explored by the Curies and others. He was a theoretician who disdained any practical applications of his discoveries. While studying the disintegration of radioactive elements, he identified the alpha particles and propounded the concept of an atom with a tiny nucleus at its center. This new concept replaced the indivisible particles of Democritus, which had dominated scientific thinking for more than two thousand years.

His greatest triumph came about 1920 when he transformed nitrogen into oxygen, and in so doing became the first person to intentionally change one chemical element into another—to cause a "nuclear reaction." Although Rutherford realized that there was a tremendous amount of energy in the atomic nucleus, he did not believe that it could ever have practical uses. Had he lived two years longer, he would have known about the discovery of uranium fission by Otto Hahn.

Lord Rutherford has been honored on stamps of New Zealand and Russia (fig. 7.5).

Perhaps the most remarkable of early twentieth-century developments was the invention and subsequent expansion of radio communication. It began with Guglielmo Marconi. Son of an Irish mother and a well-to-do Italian father, he studied physics privately under university professors. In 1894 Marconi became interested in the phenomena of Hertzian electromagnetic waves. Using Hertzian waves in conjunction with a "coherer," he was able to transform the waves into an electric current that could easily be recognized. By persistently improving his apparatus, he finally succeeded in 1901 in sending a message from England to Newfoundland. By 1905 "wireless telegraphy," that is, radio messages, had become commonplace, even though they could be used only with Morse code. Never-

Fig. 7.4. Tesla and his inventions

theless, it was a momentous invention, made more dramatic by its use in assisting ships in distress.

Marconi has been honored on several stamps, especially during 1974, the centenary of his birth. We see him portrayed twice in figure 7.6; the Marconi spark coil and the spark gap are exhibited on the United States stamp (*lower right*).

Fig. 7.5. British physicist Ernest Rutherford

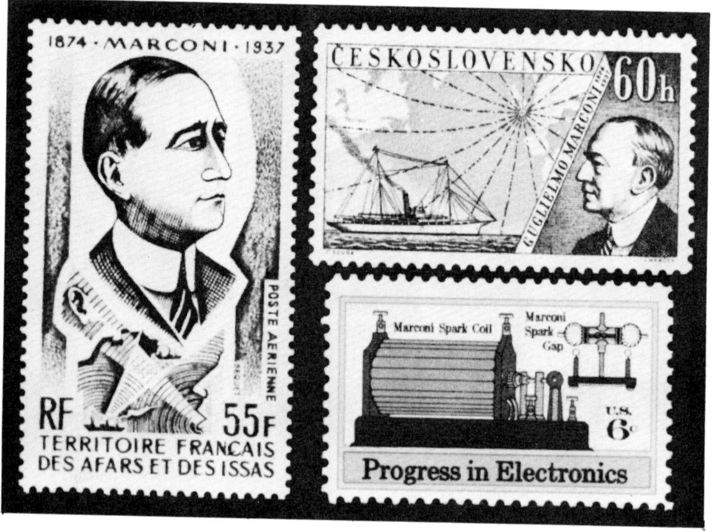

Fig. 7.6. Marconi; wireless at sea; spark coil

In 1890 Edouard Branly, professor of physics in Paris, invented the coherer used by Marconi; he is shown on the Czech stamp in figure 7.7. The Portuguese stamp above it not only shows a radio wave but epitomizes the art of electrical and electronic communication by depicting the telegraph, telephone, radio, and satellite.

Another pioneer in the development of radio, the Russian physicist Alexander Popov (1859–1905), also recognized the significance of Hertzian waves about the same time that Marconi did. By 1897 Popov was able to send a signal from ship to shore up to a distance of three miles. In time he

Fig. 7.7. Electronic communication; Branly

improved his equipment but always with an eye to studying the physics of thunderstorms and lightning, problems in which he was more interested than in the commercial applications of radio signals. He persuaded the Russian navy to install radio equipment on all its vessels. It may be said, in a sense, that radio communication was invented simultaneously and independently by Marconi and by Popov.

Portraits of Popov appear in figure 7.8: on the Bulgarian stamp (*lower left*); the Russian stamp (*top*), which depicts him demonstrating radio to Admiral Makarov; and another Russian stamp, which shows him explaining his invention of the radio.

Two more pioneers whose names are indelibly associated with radio are those of Lee De Forest and Edwin Armstrong; the contributions of both have been recognized on stamps.

Like Edison, De Forest was a prolific and imaginative inventor. In 1906 he improved Fleming's rectifier by creating the triode, the basis of the modern radio tube. This was perhaps his greatest contribution: by magnifying weak signals without distortion, he made practical not only the radio but all electronic equipment. In 1916 he was broadcasting news from a commercial radio station. By 1923 De Forest had perfected a device that made sound motion pictures possible, and half a dozen years later the "talkies" were here to stay.

Edwin Armstrong is honored on the Czechoslovakian stamp at the top

Fig. 7.8. Popov

of figure 7.9. The United States stamp below it to the left shows early forms of a radio tube, a microphone, a radio loudspeaker, and a TV camera tube. Below that another United States stamp pictures the De Forest audion, and the stamp to the right suggests the tremendous popularity of radio. The Russian stamp at the bottom depicts an early radio tube and commemorates the fiftieth anniversary of the Nizhni-Novgorod laboratory.

During World War I, Armstrong developed the so-called superheterodyne radio circuit, which not only made radio receiving sets easy to use but also enabled do-it-yourselfers to build their own sets; this added immensely to the popularity of radio. Later he developed frequency modulation (FM), which eliminated static and improved reception. Armstrong was not a good businessman, and there were frequent lawsuits between him and De Forest to establish which was the inventor of modern radio. Actually, radio may be said to have developed from the work of several inventors: surely one might well include Marconi and Popov as well as De Forest and Armstrong.

Television's practical development began about 1938 when Vladimir Zworykin, a Russian-American physicist, devised the iconoscope, a modified cathode-ray tube. By the early 1950s, TV had become a household

Fig. 7.9. Armstrong; evolution of radio

commonplace and had begun to give the radio and the movies no little competition. Many postage stamps have publicized TV over the years, including the use of transistors and solid-state circuitry.

Clockwise from the upper left in figure 7.10, a U.S. stamp depicts transistorized circuitry; a Libyan stamp issued to commemorate the inauguration of television in that country shows a symbolic television screen; a stamp of the Yemen Arab Republic depicts a television camera; and an Italian stamp shows a television screen and aerial.

Fig. 7.10. Television

THE ACHIEVEMENT OF FLIGHT

As if these breathtaking developments in electricity, physics, and electronics over a period of several decades were not exciting enough, at the same time another miracle was to be achieved. At the opening of the twentieth century, no human being had moved through the air for any significant distance unless carried aloft in a balloon or on the wings of a glider. To be sure, during the first quarter of the century, huge "powered" dirigibles were to cross the ocean on regularly scheduled passenger flights. But the heavier-than-air flying machine first became a reality about 1904, when one of the Wright brothers flew about forty kilometers (24 miles) in half an hour. A few years later the pioneer French aviator, Louis Blériot, flew across the English Channel from France to England.

In figure 7.11, the Spanish stamp at the top depicts Charles Lindbergh and his plane, the *Spirit of St. Louis*. The Paraguayan stamp under it presents a portrait of Leonardo da Vinci, foresighted inventor, with his conception of a flying machine. A stamp of the Republic of Dahomey commemorates the centennial of the birth of Louis Blériot, and the British stamp, lower left, was issued to commemorate the fiftieth anniversary of

Fig. 7.11. Aviation pioneers

the first nonstop transatlantic flight. A U.S. stamp portraying the Wright brothers, Orville and Wilbur, commemorates the forty-sixth anniversary of their first flight on 17 December 1903.

With the advent of aerial reconnaissance and aerial bombardment in World War I the airplane came of age. By 1918 the United States had established airmail service between Washington and New York, and by 1921 mail was being carried regularly across the continent by air. In 1919 two British pilots, John Alcock and A. W. Brown, negotiated the first nonstop flight across the Atlantic from Newfoundland to Ireland. However, it was Charles Lindbergh's solo flight in 1927 from New York to Paris in a little over thirty-three hours that captured the world's imagination and made modern aviation a reality. Henceforth transportation by air was to become a way of life. By the mid-1930s the familiar Douglas DC-3 had been put into regularly scheduled passenger service; its popularity was to last for some forty years.

The Swiss stamp at the top of figure 7.12 depicts a glider in symbolized aerodynamic buoyancy. The other four stamps are Russian: the two on the

Fig. 7.12. Glider; aviation pioneers Zhukovski, Chaplygin, and Mozhaisky

left portray the Russian mathematician and aerodynamics engineer N. E. Zhukovski—one with his famous lift formula and the other with a diagrammatic pressurized air tunnel. At the top right is a portrait of Sergei Chaplygin, Russian mathematician and aeronautical scientist, and below it A. F. Mozhaisky, Russian pioneer airplane builder.

As the years slipped by, planes became larger and speeds became greater. World War II led to the creation of jet-propelled planes capable of carrying hundreds of passengers at supersonic speeds of some 1250 kilometers an hour. By mid-century the jet plane had made the ocean liner obsolete, even as the automobile had already made railway passenger travel virtually obsolete. A British stamp (fig. 7.13) commemorates the first flight of the mammoth supersonic jet transport plane, the Concorde, designed to cross the Atlantic in less than four hours carrying over 130 passengers at a speed of over 2300 kilometers an hour, or twice as fast as sound. But we are getting a little ahead of our story.

Fig. 7.13. The Concorde, the world's largest supersonic jet transport

Still other events were taking place, once again in the field of physics—events that may well change completely the nature of Western civilization. To tell this story, we must go back to the first decades of the twentieth century.

RELATIVITY AND ATOMIC THEORY

Modern physics dates from about the year 1900; physics prior to that time is now generally referred to as classical physics. It was Hendrik Lorentz (1853–1928), a Dutch physicist, who first suggested that exceedingly small particles like those emitted by cathode rays contracted in size as their speed increased. He also suggested that the mass of such a particle depended on its volume: the smaller the volume, the greater its mass. Lorentz applied his conjecture to these extremely rapidly moving particles, or electrons, as they were called by Sir Joseph J. Thomson. Thus the doors of contemporary subatomic theory were already open as Planck and Einstein came on the scene.

Lorentz appears on the Dutch stamp to the left in figure 7.14. The Swedish stamp beside it portrays Thomson in the background and in the foreground Giosuè Carducci, Italian Nobel Prize winner in literature.

Fig. 7.14. Lorentz and Thomson

The German physicist Max Planck (1858–1947) reexamined the problem of blackbody radiation studied earlier by Kirchhoff, who was unable to explain the unequal distribution of the radiation frequencies. Planck succeeded in solving the problem by boldly assuming that energy could not be subdivided indefinitely but instead consisted of discrete particles, which he called *quanta*. He further assumed that the amount of energy represented by a quantum varied with the frequency of radiation. Expressed mathematically,

$$\text{amount} = h \times \text{frequency},$$

or

$$h = \frac{\text{amount}}{\text{frequency}} .$$

This constant ratio h, known as Planck's constant, plays an all-important role in quantum theory.

The constant h is featured (fig. 7.15, *center*) on the East German stamp; portraits of Planck appear on the flanking stamps.

When Planck announced his quantum theory in 1900, physicists were sceptical at first. But the theory soon found universal acceptance, and the ratio h is now regarded as a fundamental physical constant. Significantly, an entirely new branch of mathematics was developed to support his theory; it is known as *quantum mechanics*. In 1918 Planck received the Nobel Prize in physics for this truly remarkable advance in science. He

Fig. 7.15. Planck and his constant ratio *h*

had made a major contribution to the revolution in physics that Einstein was later to complete.

Because of his determined resistance to the Hilter regime, Planck found the later years of his life rather difficult despite the many honors that had been bestowed on him: he was fully recognized by the scientific community as second only to Einstein in prestige. Fortunately he survived World War II and lived to see the downfall of Nazism. He died at the age of ninety.

Albert Einstein, born in Germany in 1879 of Jewish parents, showed little promise of unusual intellectual ability as an adolescent, except in mathematics. He graduated in 1901 from a college in Switzerland, and soon found a position in the Patent Office in Berne. Here, in his spare time, he became engrossed in the study of physics. In 1905 he astounded the scientific world by publishing several startling papers, all in one year. In one of these articles, using Planck's quantum theory, he explained the so-called photoelectric effect, which classical physics could not explain. In another article he completed the theory of Brownian movement, which had previously been put forward by Maxwell and Boltzmann.

But it was the fourth paper of 1905 that shook the very foundations of physics. In this famous paper Einstein asserted that the speed of light was the fastest velocity in the universe. Furthermore, he showed mathematically why an object foreshortens in the direction in which it moves and that the faster it moves, the shorter it gets. Fantastic assertions! In brief, Einstein proposed an entirely new point of view in physics, one that was destined to revise the classical physics of Newton by introducing the concept of relativity. At first his fellow scientists did not receive the so-called "special" theory of relativity with enthusiasm, but by 1912 they embraced the new concept and proclaimed him a genius.

The inspiration for the concept of relativity was the famous Michelson-Morley experiment of 1887, in which no change was detected in the velocity of light as it changed its direction through the "ether." This negative result prompted Einstein to assume (1) that the velocity of light was constant under all circumstances, and (2) that light traveled in quanta, behaving like particles. So, without using the hypothesis of the ether, Einstein boldly assumed that there could be no such thing as "absolute motion" or "absolute rest." In other words, he asserted that all motion was relative to some specific frame of reference. On the basis of these two unprecedented assumptions, namely,

1. that the velocity of light is constant, and
2. that all motion is relative,

Einstein was able to explain the result of the Michelson-Morley experiment.

This breathtaking new viewpoint, put forward by an obscure young man of barely twenty-six, astounded the world of science and was to have an extraordinary impact on society. The idea was difficult to comprehend at first, but it soon became obvious that Newtonian physics and "common sense" were violated only when considering the vast distances in the universe and the minuscule magnitudes in the world of subatomic particles. When objects of "ordinary" size and "ordinary" velocities are contemplated, there are no measurable differences between relativity physics and Newtonian physics.

A highly significant outcome of the theory of relativity is the concept that mass is simply a particular aspect of energy—that, indeed, mass and energy are interchangeable and are related by an unbelievably simple equation:

$$E = mc^2,$$

where m represents mass and c represents the velocity of light. Since c is a comparatively large quantity, even a small mass, when multiplied by the square of the velocity, yields a tremendous amount of energy. The basic concept that mass and energy are simply two different aspects of the same phenomenon is, to be sure, very difficult to comprehend. It is, perhaps, an example of the power of mathematical symbolism to capture ideas too subtle for ordinary language symbols.

By the year 1919 Einstein's special theory of relativity had been verified in several ways, most notably by the measured deflection of light by a strong gravitational field during a particular solar eclipse. Now Einstein was world famous—he had truly become a pathfinder of the century. In 1930 he became associated with the Institute for Advanced Studies at Princeton, New Jersey. Ten years later he became an American citizen. The final years of his life were devoted, with only partial success, to searching for a general theory of relativity that would include—or unify—all gravitational and electromagnetic phenomena.

Portraits of Einstein appear on the Swiss, U.S., and Polish stamps in figure 7.16.

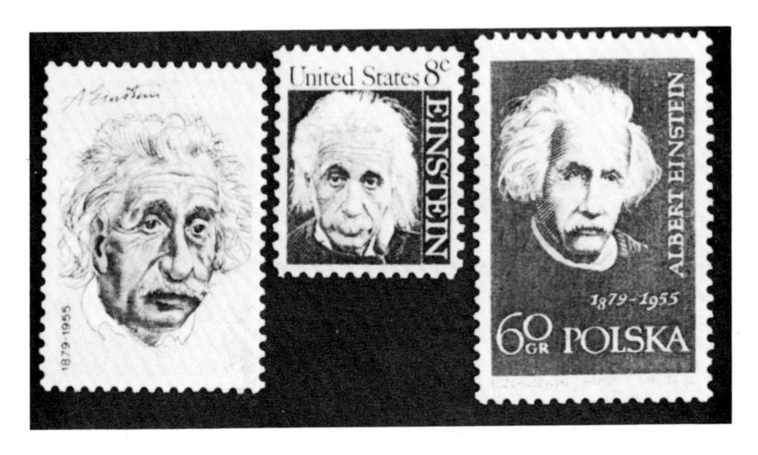

Fig. 7.16. Einstein

As World War II approached, the world was standing on the brink of catastrophe but did not know it.

You may recall that the awakening of the seventeenth century witnessed a revival of interest in atomism. By then, Newtonian mechanics and the corpuscular theory of light had assumed the ascendancy. Nevertheless, the atom was still regarded as a tiny, hard, solid object, incapable of subdivision, and indestructible. Yet about 1750, a totally unexpected hypothesis was put forward by Ruggiero Boscovich, a Serbo-Croatian mathematician and natural philosopher. Educated in Rome, Boscovich was also a poet and a priest. In mathematics he contributed to the theory of conic sections and to the theory of vortices. He is also to be remembered as the founder of modern atomism, being the first to regard an atom, not as a hard, indivisible entity, but rather as a center of force. Needless to say, his idea did not fall on fertile ground; yet by putting forth this unorthodox idea, he unknowingly foreshadowed the concept of the relativity of space and motion.

Boscovich is portrayed on the Yugoslavian and Croatian stamps in figure 7.17.

The contemporary theory of atomic structure began roughly during the first decade of the twentieth century. No single person can be credited with its development. It took the combined efforts of many scientists, chief among whom were (besides Planck and Einstein) Lorentz, Bohr, Hahn,

Fig. 7.17. Boscovich

Meitner, Pauli, de Broglie, Dirac, Schrödinger, Heisenberg, Szilard, and Fermi. The story is long and complicated, too technical to recount here in detail; a few high spots must suffice.

According to present-day atomic theory, the atom is the smallest particle of bulk matter that exhibits the unique chemical properties of an element. The atom is not a solid particle but consists mostly of empty space. At the center is the nucleus, which has a positive electrical charge and which accounts for more than 99.9 percent of the mass of the atom. Moving about this nucleus are a number of electrons, whose combined electrical charge equals that of the nucleus. The nucleus itself consists of other particles called protons and neutrons. The entire complex resembles a miniature solar system. In fact, one common method of representing atomic structure with the electrons in their orbits is seen in the familiar diagrams shown on the stamps of Sweden and the United Nations (fig. 7.18). If any one person can be considered the originator of today's concept of the atom, it would be the Danish physicist Niels Bohr (1885–1962), portrayed in the same figure on a stamp of Greenland.

We have to realize that an atom is exceedingly small. All atoms have about the same diameter, approximately a hundred-millionth of a centimeter. This is so small that the physics of the subatomic world is quite different from the observed behavior of everyday objects of ordinary size. Here again we have an example of why mathematics is an indispensable tool for describing nature.

Lord Rutherford was one of the first to suggest a model of the atom (1910); this was followed in 1913 by the early Bohr model. Then in rapid succession other models and modifications were suggested: de Broglie (1923), Pauli (1925), Schrödinger (1926), Heisenberg (1927), Dirac

Fig. 7.18. Bohr; atomic structure

(1928). Events in the world of physics were moving to a swift denouement as the curtain rose on World War II.

Yesterday and Today

1940–1975

IN MANY ways the mid–twentieth century presents a fantastic panorama. The full impact of World War II and the Korean and Vietnam wars has yet to be evaluated. These eventful decades also witnessed the creation of the United Nations and the awakening of the vast African continent.

Events in the world of science and technology were equally momentous. The famous (or infamous) insecticide DDT became available in 1942, the same year in which the first successful electronic computer made its appearance. Transistors began to be used for electronic circuits in 1953; the Salk polio vaccine was introduced in 1955; the first transatlantic telephone system was opened in 1956; the laser beam was developed in 1960; and thin-film electronic circuits had come into general use by 1962. When Crick and Watson produced their model of the helical DNA molecule in 1963, the science of genetics and molecular biology began to make tremendous strides. In 1961 the Russian astronaut Yuri Gagarin became the first human being to travel in outer space, and when American Neil Armstrong first set foot on the moon on 20 July 1969, it was indeed a dramatic moment.

Today we are in the throes of circumstances unprecedented in history: the implications of subatomic physics and nuclear power; the effects of electronic computers, automation, and datamation; and the impact of satellite telecommunication and the continued exploration of outer space.

None of these developments would have been possible without mathematics. How they will ultimately affect society, time alone can tell.

THE NUCLEAR AGE

In the fateful year 1938, Otto Hahn, a German physical chemist, found that when he bombarded uranium with neutrons, the uranium atom broke in half. This splitting of the atom, or uranium fission, was quickly confirmed by an Austrian-Swedish physicist, Lise Meitner, who at once made the discovery public. The implications of this phenomenon were immediately clear to these scientists. Fortunately, the Nazi government failed to realize the potential consequences of the discovery.

It was Niels Bohr who brought the news of uranium fission to America. The air was tense with impending drama. In December of 1942, the first nuclear chain reaction (fission of uranium isotope 235) was carried out at the University of Chicago under the direction of Arthur Compton, Enrico Fermi, and others. Further research was feverishly carried on by Bohr, as well as by Leo Szilard, a Hungarian-American physicist. Szilard had come to the United States in 1937. When he learned of the uranium fission chain reaction, he foresaw its possibilities for creating a powerful, destructive weapon. He was especially anxious to make such a bomb before Hitler's scientists did. Not without considerable effort, and aided by his friend Eugene Wigner, Szilard wrote a warning letter, which Einstein signed and sent to President Franklin D. Roosevelt; these are excerpts from that letter of 2 August 1939 addressed to "The President, White House, Washington," which was to set the Manhattan Project in motion (the complete letter can be found on p. 371 in *The Ascent of Man* by Jacob Bronowski [Boston: Little, Brown Co., 1973]; abstracts also appear on p. 262 in *Niels Bohr: The Man, His Science, and the World They Changed,* by Ruth Moore [New York: Alfred Knopf, 1966]):

> Some recent work by E. Fermi and L. Szilard, which has been communicated to me in manuscript, leads me to expect that the element uranium may be turned into a new and important source of energy in the immediate future. . . . This new phenomenon would also lead to the construction of bombs . . . and that extremely powerful bombs of a new type may thus be constructed. A single bomb of this type, carried by boat and exploded in a port, might very well destroy the whole port together with some of the surrounding territory. However, such bombs might very well prove to be too heavy for transportation by air. . . .
>
> (Signed) Albert Einstein

The atomic age was born.

Einstein is shown on the Israeli stamp in figure 8.1 (*bottom*). The Italian stamp to the right portrays Fermi and commemorates the twenty-fifth anniversary of the first nuclear chain reaction, carried out under Fermi at Chicago. The two Nicaraguan stamps to the left are from a series of ten honoring "the ten mathematical formulas that changed the face of the earth." The first shows a nuclear explosion and Einstein's famous formula, $E = mc^2$. The second commemorates de Broglie's law, $\lambda = h/mv$, which refers to "matter waves," or waves associated with particles. It states that the length (λ) of such a wave varies inversely as the product of the mass (m) of the particle and its velocity (v); the h in the numerator is Planck's constant.

Fig. 8.1. Einstein's formula; de Broglie's formula; Einstein; Fermi

An atomic bomb was created and tested at White Sands, New Mexico, on 16 July 1945. Even before the test, a group of scientists had become deeply concerned about the possible consequences of such a bomb. Szilard was among those who felt revulsion at what they had achieved. They pleaded with the authorities not to use the bomb, or if they did, not to drop it where there were people. But the military prevailed, and on 6 August 1945 an atomic bomb was exploded over the Japanese city of Hiroshima, and on 9 August a second bomb fell on Nagasaki.

The abhorrence of nuclear warfare has been depicted on several stamps (fig. 8.2).

Fig. 8.2. The abhorrence of nuclear warfare

After World War II, Szilard lost interest in nuclear physics, turned to biophysics instead, and for the rest of his life strove to ban nuclear warfare and nuclear testing. Similar feelings of guilt were held by other scientists, notably Niels Bohr, who worked unceasingly until the time of his death to further the peaceful uses of atomic energy. Albert Einstein, too, sought for more than a decade, until he died in 1955, to secure some kind of international agreement to end the threat of nuclear warfare. Twenty years later, half a dozen nations besides the United States had the capacity to wage nuclear warfare. The threat is still ominous.

Somewhere Robert McNamara has observed that "technology has now circumscribed us all with a conceivable horizon of horror that could dwarf any catastrophe that has befallen man in his more than a million years on earth."

Efforts to control nuclear weaponry and to encourage peacetime uses of atomic energy are depicted on a number of stamps (fig. 8.3). The United Nations stamp to the left commemorates the signing of a nuclear test ban treaty pledging an end to nuclear explosions in the atmosphere, outer space, and under water. A stamp of the United States (*top right*) takes note of the Atomic Energy Act, legislation sponsored by Senator Brien McMahon; the one below it asks for "atoms for peace."

Fig. 8.3. Peaceful use of atomic energy

The Atomic Energy Commission, a five-member board, was created by Congress in 1946 to promote federal and private research and development and to control the dissemination of information and the production and use of fissionable materials.

The most obvious peacetime use of atomic energy would appear to be the production of electric power by means of nuclear reactors to replace or supplement energy produced from fossil fuels or other sources. At this writing a considerable number of nuclear power plants of various types in the United States are either in operation, under construction, or in the planning stages.

In recent years many nations have pictured atomic reactors and nuclear power plants on postage stamps. A Hungarian stamp (fig. 8.4, *upper left*) pictures the Dubna Nuclear Research Institute. Moving clockwise we see a Yugoslavian stamp commemorating the first nuclear energy exposition

Fig. 8.4. Nuclear power plants

in that country. The Pakistani stamp was issued to mark the opening of the Multan Thermal Power Station, about fifty miles north of Bahawalpur, Pakistan. The Chinese stamp shows an atomic reactor and commemorates the inauguration of China's first atomic reactor and cyclotron, in Peking.

The use of nuclear power is a highly controversial question. Not the least of the problems presented by nuclear reactors are (1) the safety of their operation in view of possible accidental exposure to radioactivity, and (2) the frustrating question of how to dispose of their lethal radioactive waste products. Scientists are not completely agreed on the best means of getting rid of these waste materials, since they vary from 24 000 years to as much as 250 000 years in the length of time required for decay. Thus we are faced with an ironic situation, vividly expressed by Alvin Weinberg, one-time director of the Oak Ridge National Laboratory:

> We nuclear people have made a Faustian bargain with society. On the one hand we offer . . . an inexhaustible source of energy. . . . But the price we demand of society for this magical energy source is both a vigilance and a longevity of our social institutions that we are quite unaccustomed to. . . . Is mankind prepared to exert the eternal vigilance needed? [Alvin Weinberg, "Social Institutions and Nuclear Energy," *Science* 177 (July 1972): 33–34; also in *Smithsonian,* April 1974, p. 21]

COMPUTERS AND AUTOMATION

Today there are many kinds of computers, but generally speaking there are only two types: *analog* and *digital.* Basically, an analog computer represents the values of numbers or variables by means of physical quantities such as angular positions, linear distances, or voltages. For example, a slide rule is an analog computer; so is an automobile speedometer, which is controlled by a circular magnet actuated by a flexible shaft driven by gears at the rear of the transmission. The familiar household thermostat is another example of an analog computer: the measured temperature is converted into an electric current, which varies in magnitude as the temperature rises and falls.

Here we are concerned primarily with digital computers, which use numerals and other specific symbols. In particular, we shall discuss electronic computers, which respond only to exact signals in an electric circuit, such as "on" and "off." These machines depend basically on the binary system of notation, where all numbers can be represented by using only the symbols 0 and 1. Electronic computers also use specific symbols, which represent arithmetical or other operations to be performed. These operational symbols are assigned numerals, which can then be fed into the computer. The information given to a computer is first punched on cards or on magnetic tape.

Two East German stamps in figure 8.5 picture typical electronic computers.

Fig. 8.5. Electronic computers

The basic functions of a digital computer are input, storage, control, processing, and output. The computer is capable of adding, subtracting, multiplying, dividing, and listing. But beyond these, it can also make decisions in accordance with stored instructions; preparing these instructions is known as programming. A computer system usually consists of an assembly of electronic and electromechanical components: the computer itself, storage and memory devices, switching circuits, magnetic tapes, printout devices, and so on. The speed with which computation takes place and information is retrieved is phenomenal—on the order of a millionth a second.

The stamps in figure 8.6 picture familiar aspects of electronic computers. The Canadian stamp (*upper right*) exhibits punched paper tape and magnetic tape, used in conjunction with census statistics. The Dutch stamp below it and the Norwegian stamp (*upper left*) represent the typical punched card. The Australian stamp depicts computer circuitry against the background of a conventional abacus, and the one from Israel represents a punched card and magnetic tapes.

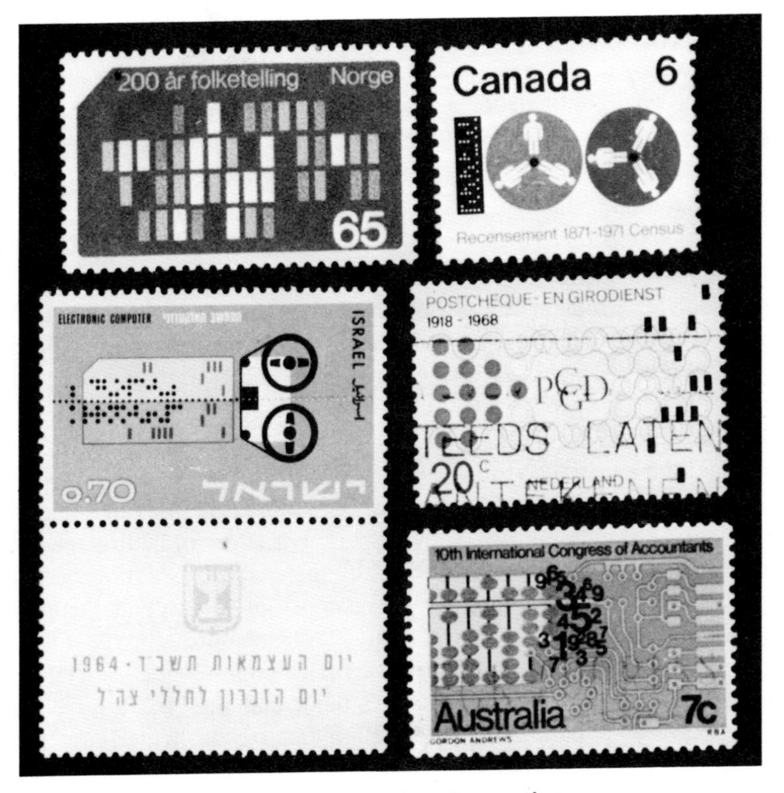

Fig. 8.6. Punched cards and magnetic tape

The first all-purpose, all-electronic digital computer, the ENIAC, was completed in 1946. It could carry out in a few hours computations that would take scores of scientists a year or more to complete. A few years later, the mathematician John von Neumann (1903–1957) and his colleagues extended the potentialities of electronic computers enormously. Today they have become veritable supermachines.

There is no denying that the modern electronic computer is a powerful tool that is shaping the world of today as well as that of tomorrow. To be sure, the computer merely does arithmetic, that is, performs operations with numerals. But when one remembers that any desired meaning can be assigned to arbitrary symbols and that these symbols can be associated with numerals, it becomes clear that interrelationships between the symbols can be discovered by manipulating the symbols. The computer has vast powers to manipulate symbols according to a set of rules. It is this power that makes the computer admirably adaptable for tasks that require large masses of data to be processed and refined in a reasonably short period of time. And as we have seen, the speed with which the computer operates is virtually incomprehensible to the human mind.

The many uses to which the computer has already been put include a wide range of tasks in mathematics, industry, technology and engineering, medicine, agriculture, meteorology, economics, and the sociological sciences.

In mathematics computers have been used to furnish proofs in the theory of numbers, to prove theorems in queuing theory, to solve problems in crystallography and metallurgy, to extend our knowledge of celestial mechanics, and to track the courses of rockets and satellites in outer space.

In the field of recreational mathematics, the computer has been of great help in solving problems involving prime numbers, perfect numbers, networks, "squaring the square" dissections, tournament and other combinatorial problems (such as Instant Insanity), game strategies, and many others.

In industry and technology, computers are used in either of two ways: (1) to analyze data *after* it has been collected from an experiment or from other experiences, or (2) to perform an experiment or try out some other experience by simulation *before* the experiment is performed or the operation is carried out. The former is generally referred to as "data processing," the latter as "simulation methods." Examples of the second method are the flying (on paper) of thousands of trajectories before a bomb is actually constructed or an airfoil is designed, the calculating of stresses in the steel framework of a skyscraper before it is built, and the "sea testing" of an atomic-powered submarine by simulation before the actual construction and launching of the vessel.

Automation plays a role in almost every area of human endeavor, and the varied applications of computers and data-processing equipment are revealed on many stamps. For example, in figure 8.7, one of the East German stamps (*top left*) shows a fully automated offset printing press, and the other depicts an automated drill and milling machine with instrument panels visible on the drill. The stamp of the Ivory Coast (*lower right*), issued to publicize the development of data processing, shows a computer operator using a keypunch machine. Above it, the Tunisian stamp, issued to publicize the introduction of electronic equipment into the postal service for automatic letter sorting, depicts a "humanized computer."

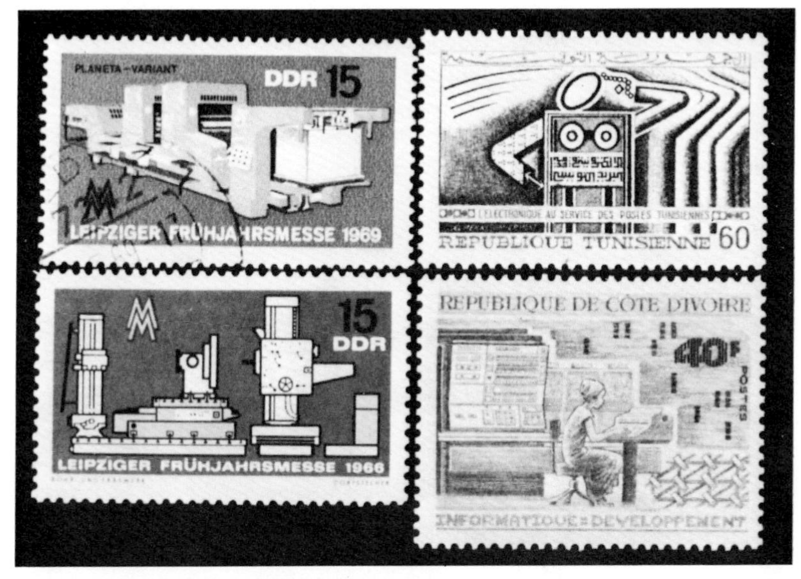

Fig. 8.7. Automatic machinery and data processing

The endless variety of seemingly unrelated applications of the computer stagger the imagination. Computers can play chess, translate languages, and create musical compositions. In everyday affairs they help us in making airplane reservations, buying theater tickets, controlling traffic in metropolitan areas, and routing thousands of freight cars daily. In one huge modern freight yard, a radar device registers all the factors that affect a freight car's rolling down a gentle incline—weight, size, speed, wind, weather, and so on—and an electronic computer regulates the retarder brake in the track under the moving car so it will couple safely and gently to another car standing far down in the yard below.

In wartime, computers control the distribution of strategic materials, the production of military equipment, the management of the work force and troop movements, and many other matters. In peacetime, computer techniques are of great value in such diverse activities as taking the national census, processing millions of social security checks, finding oil deposits in underground strata, predicting weather with the aid of satellites, helping farmers make plans for better use of land, reading electrocardiograms, and monitoring coronary patients. These are only a few of the many applications of computers and electronic data-processing equipment that are profoundly affecting our culture.

Of interest are some additional uses of computerized equipment as revealed on stamps. In figure 8.8 we see on the Australian stamp a soil sample being tested chemically with the aid of a computer and punched paper tape; the stamp commemorates the ninth International Soil Science Congress at the University of Adelaide. A Russian commemorative (*lower left*) publicizes electronic mechanization and automation in industry generally. The other Russian stamp (*lower right*), showing a computer tape, a cogwheel, and the atomic symbol on a factory background, is further evidence of the computerization of industry. The East German stamp (*upper right*) pictures a computerized threshing and silage-producing machine.

Fig. 8.8. Applications of computers

Many industries are now partially or almost wholly automated by means of electronic equipment and computers: steel mills, glass-manufacturing plants, oil refineries, chemical factories, paper-making plants, textile mills, and so on. For example, in one factory that produces synthetic fibers and textiles, even the dyeing process is carefully programmed and computerized. An Albanian stamp (fig. 8.9, *top*) reveals electronic technicians in a Chinese steel mill operating a computerized control center to pour a ladle of molten metal; the East German stamp below it shows operators at an electronic control panel in a petrochemical plant.

Fig. 8.9. Electronic mechanization of industry

As a final instance of ingenious applications of computers we note the solving of a fascinating archaeological problem. Recently a distinguished Egyptologist was able to reconstruct the temple of Queen Nefertiti. Descriptive data on more than thirty-five thousand stone blocks from the ancient shrine were fed into a computer to determine the exact position of each stone in the original structure, much as the pieces of a jigsaw puzzle are fitted together.

In addition to the fabulous variety of tasks for which computers have been adapted, these machines are bringing about new methods of learning and knowing—a great adventure for the human mind. And all this devolves from the mathematics of the binary system of two symbols, 0 and 1.

THE CHALLENGE OF OUTER SPACE

Unlike a jet engine, which can function only where the atmosphere is dense enough to support combustion, a rocket is a reaction-propelled body that carries along not only its fuel but all the oxygen it needs for combustion. Since a rocket travels best where there is no atmosphere, it is well suited for space exploration, meteorological studies, satellite telecommunication, and related purposes.

Although an interest in rockets has existed since early times, the architect of modern rocketry is an American, Robert H. Goddard, who, in effect, inaugurated the space age. Even as a young man he dreamed of interplanetary travel, and in 1909 he made theoretical studies of liquid fuels for rockets. In 1914 he became professor of physics at Clark University in Worcester, Massachusetts, where he remained for thirty years. In 1923 Goddard built the first rocket engine, which used gasoline and liquid oxygen. In 1930 he moved to a remote place in New Mexico, where he expanded his experimental work, producing rockets that attained speeds of 887 kilometers an hour and ascended to a height of nearly 2.5 kilometers. Until his death in 1945 he continued his research with unabated zeal: his goal—even higher altitudes and greater speeds.

A gifted and imaginative inventor, Goddard was encouraged at first by financial support from the Smithsonian Institution and later from the Guggenheim Fund for the Promotion of Aeronautics. Throughout his career he was granted more than two hundred patents. During World War II he offered his services to the United States military but met with a cool reception; he died in 1945 without seeing his country enter the space age. Our government has never fully acknowledged its debt to his achievements.

Goddard is shown on an American stamp (fig. 8.10), which also pictures an Atlas rocket and a launching tower at the Cape Kennedy Space Center, located at Cape Canaveral, Florida. The French stamp below it was issued to honor another aviation and space expert, Robert Esnault-Pelterie; it depicts a Diamant rocket and an A-1 satellite.

Meanwhile, in Russia, rocket research had begun in the 1920s under the guidance of Konstantin Tsiolkovsky (1857–1935), an able scientist and mathematician and an expert in aeronautics and astronautics. One of the first to study the aerodynamics of an airfoil by using a wind tunnel, he pioneered in developing the theory of rocket travel in space.

Tsiolkovsky is portrayed on two Russian stamps in figure 8.11. The Polish stamp (*center*) pictures Tsiolkovsky's rocket and gives his formula for rocket speed: $v = w \log \frac{m_0}{m}$. Here v presents the velocity of the rocket with respect to the earth; w, the jet velocity in relation to the rocket; m_0, the initial mass of the rocket; and m, its final mass.

The young Tsiolkovsky, who was born some twenty-five years earlier than Goddard, lost his hearing at the age of nine as a result of scarlet

Fig. 8.10. Pioneers of rocketry: Goddard and Esnault-Pelterie

fever. Nevertheless, he assiduously studied chemistry, mechanics, mathematics, and astronomy. During the first fifteen years of the twentieth century he was beset by personal tragedies; furthermore, his early researches in aerodynamics during those few years were unappreciated by the Russian government. However, after the revolution of 1917, he spent the next eighteen years in rewarding research in stratospheric exploration and interplanetary flight—this time with the blessing of the new regime in

Fig. 8.11. Tsiolkovsky and his formula

Russia. By the early twenties the Soviet Union had well-established rocket research centers in Moscow, Leningrad, and Kazan.

Here are three more stamps honoring Tsiolkovsky (fig. 8.12): the one at the top shows his formula for rocket speed; the Bulgarian stamp (*lower left*) shows his portrait beside a rocket launching; the Russian stamp commemorates the centennial of his birth.

Fig. 8.12. Tsiolkovsky: his formula and his rockets

The Germans, too, had made some progress in military rockets, having successfully dropped about two thousand V-2 rockets on London and Antwerp in 1944–45. Most of what they knew about rockets they had quietly learned from the work of Goddard.

In astronomy, any small planetlike body that revolves in an orbit about a given planet is known as a *satellite*—for example, the moon is a satellite of the earth. An artificial satellite is any device designed to be propelled far enough from the earth's surface for it to move in an orbit about the earth. Such satellites are usually provided with instruments to record automatically many kinds of data, such as ultraviolet and X-ray radiation from the sun, the earth's magnetic field at high altitudes, the presence of cosmic rays, the concentration of meteoric dust, and other valuable information.

The first artificial satellite ever to circle the earth was launched by the Russians on 4 October 1957. Known as *Sputnik I,* it was a sphere a little less than sixty centimeters in diameter and weighing about eighty kilograms. Its initial altitude was about 900 kilometers above the earth, and it revolved with an orbital velocity of nearly 9 kilometers a second, or about 18 000 miles an hour. It made one revolution around the earth about every hour and a half. It also started a revolution of another kind: it initiated the space age and challenged the United States.

The challenge was soon answered. In January 1958, the United States launched the satellite *Explorer I* at Cape Canaveral, Florida; it was a cylindrical rocket about two meters long and fifteen centimeters in diameter, weighing nearly fourteen kilograms. Shortly thereafter, *Explorer III,* carrying five kilograms of instruments, was orbited with a velocity of 28 800 kilometers an hour. Between January and July of 1958 the United States launched four Explorers (one unsuccessful) as well as *Vanguard I.*

The drama of the sputniks is recorded on stamps. The Romanian set at the top of figure 8.13 pictures *Sputnik I* over the earth (*left*) and *Sputnik I* and *II* circling the globe (*right*). The Polish stamp at the bottom pictures *Sputnik III.*

During the next few years, these early satellites, both American and Russian, were to lead to the development of an amazing complex of

Fig. 8.13. *Sputnik I, II,* and *III*

telecommunication equipment enabling people to witness events on their television screens *as they were happening* halfway around the earth. At first, it was stranger than science fiction; today, it is commonplace. Such is the fantastic pace of contemporary technical achievement.

Communication satellites traveling in earth orbits provide links between different places on the earth's surface. Such satellites, developed since World War II, constitute the most significant nonmilitary achievement in space technology thus far. They make possible the interchange of live television broadcasts between widely separated nations on all continents. These satellite communication systems also make use of international telephone service as well as microwave radio relay systems and sundry electronic equipment.

Relay stations are located in more than fifty countries. The desired signals are transmitted from one station to an orbiting satellite; the equipment aboard the satellite amplifies the signals and rebroadcasts them to another station. Among the well-known satellite systems are the several INTELSAT systems, satellites weighing from about 40 kilograms (the *Early Bird),* up to about 680 kilograms, the weight of *INTELSAT IV,* which has a capacity for twelve one-way television channels or about three thousand to nine thousand two-way telephone channels.

Many of these achievements are reflected on stamps. On the Colombian stamp (fig. 8.14) we see the *Early Bird* satellite, the Southern Cross, and radar equipment. The Republic of the Congo depicts the Breguet dial telegraph, the ITU emblem, and a TELSTAR satellite on one of its stamps. The French stamp commemorates the launching of the *D1* satellite at Hammaguir on 17 February 1966, and the Nigerian one exhibits the United States *Mercury* capsule over the Kano tracking station in Nigeria.

In addition to military uses and telecommunications, satellites serve astronomy by carrying telescopes above the earth, supplementing the use of radio telescopes. As we saw in an earlier chapter, the radio telescope is a parabolic reflector antenna (or dish) that receives extraterrestrial radio noise from outer space. Although space astronomy is still a young science, highly significant results have already been achieved. Important observations have been made from just above the earth's atmosphere; instruments have been landed on the moon, Venus, and Mars; the side of the moon facing away from the earth has been carefully photographed; a rocket flight near Mars has provided photographs of the surface of that planet; and in recent years a number of orbiting solar and astronomical space observatories have been operating successfully. Not the least rewarding aspect of these developments is the increased trend toward international cooperation: observatories and telescopes have been established in South Africa, Australia, Argentina, and Chile, as well as in other places.

Fig. 8.14. Satellites in space

These matters have been reflected from time to time on postage-stamp designs. Figure 8.15 shows a Chilean stamp issued to recognize the inauguration of ENTEL Chile, the first commercial satellite communications ground station, located at Longovilo, Chile. The stamp of the Malagasy Republic pictures the Philibert Tsiranana Radar Station. The Canadian stamp shows the Canadian satellite *Alouette II* orbiting the earth; it was launched in California in 1965 as part of the Canadian-American program of space research.

Although long-distance electronic communication by means of satellites still depends on radio waves, it is not inconceivable that in the near future the use of laser beams to send messages to and from satellites would make available a multitude of channels: thus there could be an individual television channel for each of us!

As might have been expected, the spectacular success of military rockets and communication satellites soon gave rise to more ambitious and imaginative adventures in exploring outer space. Fortunately, electronic digital computers were by now highly developed, for without them further advances in space exploration would have been impossible. We mentioned

Fig. 8.15. Telecommunication by satellite

the "three-body problem" in an earlier chapter: the equations describing the relative motions of the earth, a spacecraft, and the moon are extremely complicated. The high-speed computer integrates these equations numerically while the spacecraft is in motion. On command, the computer furnishes a direct readout of the spacecraft's motion and makes it possible at any instant to compare the actual flight path to the planned path.

By 1959 the exploration of space with unmanned satellites had begun in earnest, simultaneously and independently, by both the United States and the Soviet Union. In 1965 France launched the *Diamant,* its first satellite, becoming the third nation to enter the arena. The support of Congress has enabled the National Aeronautics and Space Administration (NASA), established in 1958, to pursue these explorations on a dramatic, comprehensive scale. The list of flights and space probes became more and more extensive during the decade of the sixties and beyond. Names that have since become household words include Vanguard, Explorer, Luna, Pioneer, Echo, Mariner, and Ranger. Such spacecraft have been pictured on many postage stamps of nearly every nation, including most of the countries of Asia.

As early as 1961 both the United States and the Soviet Union began launching *manned* space flights. In April of that year the Russian astronaut Yuri Gagarin became the first to orbit in outer space. He was followed by American astronauts Alan Shepard, Jr., in May 1961 and Virgil Grissom in July 1961. In 1962 John H. Glenn, Jr., and M. Scott Carpenter made similar flights in space. The race between the United States and Russia was on. The climax came in 1969 when Neil A. Armstrong, Edwin E. Aldrin, Jr., and Michael Collins took *Apollo II* to the moon. On 20 July when Armstrong stepped from the spaceship and set foot on the moon's surface, his first words were, "One small step for a man, one giant step for mankind!" It was more than that: the dream of centuries had come true.

The story of the conquest of space is a fascinating one. It has been told on many hundreds of stamps, but there is room here to exhibit only a few. The first of these United States stamps (fig. 8.16) depicts Project Mercury, the *Friendship 7* capsule commanded by John Glenn. The stamps below it are a se tenant pair, one showing an astronaut walking in space, the other, a picture of the *Gemini 4* capsule and the edge of the earth; they were issued to commemorate U.S. accomplishments in space.

Fig. 8.16. U.S. space pioneers

In 1973 the United States initiated Skylab, a program consisting of a series of three-man flights to conduct experiments in a space workshop orbiting the earth. The workshop, depicted in figure 8.17, is a cylinder about 10.8 meters long and 6.6 meters in diameter. It can accommodate not only three astronauts but also ample equipment for tests and investigations. Three other U.S. stamps complete the figure: the one showing the

Fig. 8.17. Moon landing

surface of the earth and the moon commemorates the *Apollo 8* mission, which put the first person in orbit around the moon 21–27 December 1968; another commemorates the first landing of human beings on the moon; another, depicting the Moon Rover, pays tribute to a decade of United States achievement in space.

As we look back on the developments in mathematics, science, and technology since 1900, we can only marvel at the changes that have taken place in the span of one lifetime. People who might have seen one of the first airplanes flying overhead in the early days of the twentieth century could have lived to see, on a small screen in their own living room, two men walking on the moon.

What miracle can we expect next?

Epilogue

AND so we end our story. To be sure, there are a number of gaps where no stamps exist to commemorate some outstanding mathematician or scientist. But the main thread of the story is clear: how mathematics evolved, some of the people who created or developed it, and the relation of mathematics to science and technology.

As you were reading, you may have noted the interplay between mathematics and science, especially since the Renaissance. From the sixteenth through part of the eighteenth century this fruitful relationship was concerned largely with astronomy, navigation, surveying, and mechanics. During the nineteenth and twentieth centuries the emphasis shifted to physics—chiefly problems about light, magnetism, electricity, atomic theory, and electronics.

Some forty years ago Sir James Jeans remarked that "*all* the pictures which science draws of Nature, and which alone seem capable of according with observational facts, are *mathematical pictures*." We need only think of the paths of the planets in outer space, the symmetry of a snowflake, the geometry of crystals, the economy of the honeycomb, the spiral of the snail's shell, the strands of a spider's web, the efficient arrangement of leaves on a stem, the exponential rate at which radioactive material decays, the periodicity of the heartbeat, and countless other natural phenomena.

At the same time, we must realize that many of our artifacts and inventions reveal their dependence on mathematical properties. Consider the wheel and the lever, the dignity of the capital of a Greek column, the majesty of the Gothic arch, the glorious dome of Saint Peter's, the flying buttresses of a medieval cathedral, the graceful sweep of the cables of a suspension bridge, the gleaming spherical storage tanks of an oil refinery, the parabolic reflector of an outdoor band shell or a radar tracking station, the exquisite patterns of sixteenth-century Italian lace, the naive designs found in Mexican pottery or Indian rugs, or the austerity of a skyscraper towering upward for nearly a quarter of a mile.

Let us look a little further at the intimate connection between science and mathematics by examining briefly the nature of contemporary mathematics, humanity's greatest invention. Pure mathematics today is a complex of many mathematical systems, each highly abstract and each built upon a particular set of arbitrary assumptions. Using specialized symbolism and dealing with generalizations, the mathematician employs imaginative "models" (not physical devices) to interpret the subject matter of any given mathematical system. The essence of the procedure has been aptly expressed by an American physicist, W. F. G. Swann:

> The pure mathematician . . . will set up a branch of mathematics founded upon certain postulates having to do with quantities, letters, etc., that he chooses to be talking about. In this mathematical scheme, there will appear relationships between certain quantities which occur in the mathematics, and it will be his hope to invent a scheme of mathematics of this kind which shall form an analogue of the regularities of nature in the sense that there may be a one-to-one correspondence between certain things in the mathematics and the observable phenomena in nature. . . . When the correspondence has been set up, the postulates of his mathematics become the laws of nature in the physics. [W. F. G. Swann, "Reality in Physics," *Science* 75 (1932): 113–14]

In this way it becomes possible to apply mathematical thinking with considerable success not only to the physical sciences and technology but also to medicine, economics, history, sociology, and business. In particular, witness the revolution in physics since 1900. Without the necessary mathematics there would be no radio or television, no computers or automation, no telecommunication or space program. Truly, mathematics is the "handmaiden of the sciences" as well as a unique art in its own right. It always has been.

Appendix A

A Checklist of Stamps—Mathematicians and Scientists

The mathematicians and scientists who are listed here have been commemorated on stamps for their contributions to knowledge. Such a list can clearly never be "complete": new stamps are constantly being issued. For instance, in 1973–74 more than fifty nations issued stamps honoring Copernicus; in 1975, a plethora of stamps on the metric system appeared; and in 1976–77 there was a deluge of stamps commemorating Alexander Graham Bell, the inventor of the telephone.

The names of the mathematicians and scientists and their dates are followed by the names of the countries that have issued stamps in their honor. These abbreviations were used:

China = People's Republic of China
D.D.R. = German Democratic Republic (East Germany)
Germany = German Federal Republic (West Germany)
Russia = Union of Soviet Socialist Republics
Tanzania = Tanzania-Kenya-Uganda
U.A.R. = United Arab Republic
U.N. = United Nations
U.S.A. = United States of America
Y.A.R. = Yemen Arab Republic

The Scott number follows the name of the issuing country. Stamps not recognized by Scott are designated "unlisted." The date of issue is given in parentheses after the Scott number.

Abel, Niels Henrik, *1802–1829*
Norway, 145–48 (1929)

"al-" names. *See following element*

Alembert, Jean d', *1717–1783*
France B332 (1959)

Alhazen, *965–1039*
Pakistan 281 (1969)
Qatar 235 (1971)

Ampère, André M., *1775–1836*
Afars and Issas C90 (1975)
France 306 (1936); 626 (1949)
Mali C258 (1975)
Monaco 1001 (1975)

Archimedes, *ca. 287–212* B.C.
D.D.R. 1501 (1973)
Nicaragua C751 (1971)
Spain 1159 (1963)

Aristotle, *384–322* B.C.
Greece RA91 (1956)

Armero, Julio, *1865–1920*
Colombia 573 (1949)

Armstrong, Edwin, *1890–1954*
Czechoslovakia 954 (1959)

Augustine of Hippo, *354–430*
Algeria 261 (1954)

Avicenna, *980–1037*
Afghanistan 390, 391 (1951); B1, B2 (1952)
D.D.R. 106 (1952)
Lebanon 223, 224 (1948)
Persia (Iran) 1226, 1227 (1962); B31–B35 (1954)
Poland 558 (1952)
Qatar 237 (1971)
Syria C340 (1965)

Baily, Francis, *1774–1844*
Great Britain 616 (1970)

Bell, Alexander Graham, *1847–1922*
Canada 274 (1947)
Niger Republic C191 (1972)
U.S.A. 893 (1940)

Bjerknes, Vilhelm, *1862–1951*
Norway 403, 404 (1962)

Bohr, Niels, *1885–1962*
Denmark 409, 410 (1963)
Greenland 57, 58 (1963)

Bolyai, Farkas, *1775–1856*
Hungary 479 (1932); 2347 (1975)

Bolyai, János, *1802–1860*
Hungary 1321 (1960)
Romania 1345 (1960)

Boltzmann, Ludwig, *1844–1906*
Nicaragua C749 (1971)

Boscovich, Ruggiero, *1711–1787*
Croatia 59, 60 (1913)
Yugoslavia 595 (1960)

Bouguer, Pierre, *1698–1758*
Ecuador 347, 349, 351, C39, C41 (1936)

Brad, Ion Ionescu de la, *1818–1891*
Romania 2012 (1968)

Brahe, Tycho, *1546–1601*
Denmark 300 (1946)
Y.A.R. (unlisted)

Branly, Edouard, *1844–1940*
Czechoslovakia 951 (1959)
France 471 (1944); B438 (1970)
Monaco 615 (1965)

Broglie, Louis de, *1892–*
Nicaragua C750 (1971)

Broglie, Maurice de, *1875–1960*
France B439 (1970)

Buffon, Comte de, *1707–1788*
France B241 (1949)

Carmona, Antonio, *1869–1951*
Portugal 556 (1934); 650–57 (1945)

Carnot, Lazare N. M., *1753–1823*
France B251 (1950)

Carnot, Sadi, *1837–1894*
France B287 (1954)

Chang Heng, *78–139*
China 245 (1955)

Ch'ang Sui, *683–727*
China 247 (1955)

Chaplygin, Sergei, *1869–1944*
Russia 945, 946 (1944)

Chebyshëv, Pafnuti L., *1821–1894*
Russia 1050, 1051 (1946)

Cierva, Juan de la, *1896–1936*
Spain C100–C108 (1939); C109–C116 (1941–47)

Comte, August, *1798–1857*
Brazil 854 (1957)
Bulgaria 1001 (1958)
France 848 (1957)
Romania 1218 (1958)

Constant, Benjamin (Botelho de Magalhães), *1833–1891*
Brazil 175 (1906); 484 (1939); 807 (1954)

Copernicus, Nicolaus, *1473–1543*
Afars and Issas C83 (1973)
Ajman M1059 (1971)
Albania 1481–86 (1973)
Brazil 1301 (1973)
Bulgaria 2086 (1973)
Burundi 431–34, C183–C186 (1973)
Chad C109 (1973)
China 205 (1953)
Colombia C593 (1974)
Comoro Islands C56 (1973)
Congo People's Republic C160 (1973)
Dahomey C185, C186 (1973)
D.D.R. 1461 (1973)
France 857 (1957)
French Polynesia C95 (1973)
Germany 1104 (1973)
Hungary 2218 (1973)
India 537 (1973)
Libya 489, 490 (1973)
Mali C178 (1973)
Mexico C416 (1973)
Mongolia 723–26 (1973)
Pakistan 336 (1973)
Paraguay (unlisted)
Poland 192, 194 (1923); N60, N61 (1940–41); NB1 (1940); NB23 (1942); NB 27 (1943); 360 (1945); 515 (1951); 578, 579 (1953); 672 (1955); 885 (1959); 982 (1961); 1229 (1964); 1659–61 (1969); 1745–47 (1970); 1818–21 (1971); 1915–18, B127 (1972); 1944–45 (1972); 1956–60, 1979, 1982–85, B128, B129 (1973)
Romania 2405 (1973)
Russia 1752 (1955); 4060 (1973)
Rwandaise 565–70 (1973)
Togo Republic 843, C201 (1973)
U.S.A. 1488 (1973)
Uruguay 870 (1973)
Vatican City 436 (1966); 537–40 (1973)
Venezuela 1013–15 (1973)
Y.A.R. (unlisted)

Coulomb, Charles Augustin de, *1736–1806*
France B352 (1961)

Cristescu, Vasile, *ca. 1930*
Romania 596 (1945)

Cusanus, Nicolaus, *1401–1464*
Germany 792 (1958)
Vatican City 395, 396 (1964)

d'Alembert, Jean. *See* Alembert

da Vinci, Leonardo. *See* Vinci

de Broglie. *See* Broglie

Democritus, *ca. 460–370* B.C.
Greece 717 (1961)

Descartes, René, *1596–1650*
France 330, 331 (1937)

De Witt, Jan, *1625–1672*
Netherlands B177 (1947)

Dürer, Albrecht, *1471–1528*
Ajman (unlisted)
D.D.R. 1298 (1971)
Germany 362 (1926–27); 827 (1961–64)
Germany (Berlin) 9N179 (1961–62)
Niger Republic C68 (1967)
Romania 2292 (1971)
Rwanda 430 (1971)

Einstein, Albert, *1879–1955*
- Argentina 951 (1971)
- Ghana 190 (1964)
- Israel 117 (1956)
- Mali C250 (1975)
- Nicaragua 878 (1971)
- Paraguay (unlisted)
- Poland 882 (1959)
- Switzerland 549 (1972)
- U.S.A. 1285 (1966)

Eötvös, Loránd, *1848–1919*
- Hungary 471 (1932); 840 (1948)

Euler, Leonhard, *1707–1783*
- D.D.R. 58 (1950); 353 (1957)
- Russia 1932 (1957)
- Switzerland B267 (1957)

Farabi, al-, *870–950*
- Persia (Iran) 947, 948 (1951)
- Qatar 234 (1971)
- Turkey 1037–40 (1950)

Fermi, Enrico, *1901–1954*
- Italy 976 (1967)

Foucault, Jean Bernard Léon, *1819–1868*
- France 871 (1958)

Franklin, Benjamin, *1706–1790*
- U.S.A. 1073 (1956)

Galilei, Galileo, *1564–1642*
- Ascension 141 (1971)
- Burundi Republic 295 (1969)
- Czechoslovakia 1231 (1964)
- Ecuador (unlisted)
- Hungary 1592 (1964)
- Italy 419–22 (1942); 888, 889 (1964); D16 (1933); D18 (1945)
- Mexico C378 (1971)
- Niger Republic C125, C130 (1970)
- Paraguay (unlisted)
- Romania 1647 (1964)
- Russia 2986 (1964)
- Y.A.R. (unlisted)

Galileo. *See* Galilei

Gauss, Carl Friedrich, *1777–1855*
- D.D.R. 1811 (1977)
- Germany 725 (1955); 1246 (1977)

Gerbert (Pope Sylvester II), *ca. 940–1003*
- France B384 (1964)
- Hungary 511, 516 (1938)

Goddard, Robert H., *1882–1945*
- U.S.A. C69 (1964)

Godin, Louis, *1704–1760*
- Ecuador 347, 349, 351, C39, C41 (1936)

Guericke, Otto von, *1602–1686*
- Germany 472 (1936)

Guldberg, Cato Maximilian, *1836–1902*
- Norway 452, 453 (1964)

Gusmão, Bartholomeu de, *1675–1724*
- Brazil C22 (1929–30)

Hamilton, William Rowan, *1805–1865*
- Eire 126, 127 (1943)

Hell, Maximilian, *1720–1790*
- Czechoslovakia 1670 (1970)

Helmholtz, Hermann L. F. von, *1821–1894*
- D.D.R. 62 (1950)

Herschel, William, *1738–1822,* and John Herschel, *1792–1871*
- Great Britain 616 (1970)

Hertz, Heinrich Rudolph, *1857–1894*
- Czechoslovakia 953 (1959)
- D.D.R. 354 (1957)
- Germany 762 (1957)
- Mexico C332 (1967)

Hipparchus, *ca. 190–126* B.C.
- Greece 835 (1965)

Hronec, Juraj, *1881–1959*
- Czechoslovakia 1103 (1962)

Huygens, Christian, *1629–1695*
- Netherlands B36 (1928)

ibn-al-Haytham. *See* Alhazen

Idachimescu, A. O., *ca. 1925*
Romania 596 (1945)

Imhotep, *ca. 3000* B.C.
Egypt 153 (1928)

Ionescu, Ion, *ca. 1925*
Romania 596 (1945)

Juan, Jorge, *1712–1774*
Ecuador 348, 350, C40 (1936)

Jungius, Joachim, *1587–1657*
D.D.R. 352 (1957)

Kant, Immanuel, *1724–1804*
D.D.R. 1542 (1974)
Germany 356 (1926–27); 364 (1927); 831 (1961–64)
Germany (Berlin) 9N183 (1961–62)
Haiti 414, C105–C107a (1956)

Kepler, Johannes, *1571–1630*
Austria B282 (1953)
Dahomey C142, C143 (1971)
D.D.R. 1275 (1971)
Fujeira (unlisted)
Mexico C379 (1971)
Romania 2309 (1971)
Y.A.R. (unlisted)

Kindi, Abu-Usuf al-, *ca. 813–873*
Iraq 303 (1962)

Kirchhoff, Gustav Robert, *1824–1887*
D.D.R. 1541 (1974)
Germany (Berlin) 9N345 (1974)

Kovalevski, Sonya, *1850–1891*
Russia 1570 (1951)

Krylov, Aleksei N., *1863–1945*
Russia 1792 (1956); 2713 (1963)

Kucera, Oton, *1857–1931*
Yugoslavia 493 (1957)

La Condamine, Charles M. de, *1701–1774*
Ecuador 347–51, C39–C42 (1936)

Lagrange, Joseph Louis, *1736–1813*
France 869 (1958)

Langevin, Paul, *1872–1946*
France 608 (1948)

Laplace, Pierre Simon de, *1749–1827*
France B298 (1955)

Leibniz, Gottfried Wilhelm von, *1646–1716*
D.D.R. 66 (1950)
Germany 360 (1926–27); 962 (1966)
Romania 1855 (1966)

Leonardo da Vinci. *See* Vinci

Leverrier, Urbain Jean Joseph, *1811–1877*
France 870 (1958)

Liapunov, A. M., *1857–1918*
Russia 1951 (1957)

Lobachevski, Nikolai Ivanovich, *1793–1856*
Russia 1575 (1951); 1822 (1956)

Lorentz, Hendrik Antoon, *1853–1928*
Netherlands B35 (1928)
Sweden 618 (1962)

Maimonides (Rabbi Moses ben Maimon), *1135–1204*
Grenada 401 (1971)
Israel 74 (1953)
Spain 1463 (1967)

Maldonado, Pedro V., *1707–1748*
Ecuador C42 (1936)

Marconi, Guglielmo, *1874–1937*
Afars and Issas C87 (1974)
Canada 654 (1974)
Colombia 829 (1975)
Czechoslovakia 952 (1959)
India 646 (1974)
Italy 397–99 (1938); 909 (1964); 1141, 1142 (1974)
Rwandaise 587–92 (1974)

Maróthy, Gijorgy, *1715–1744*
Hungary 533 (1938)

Maxwell, James Clerk, *1831–1879*
Mexico C332 (1967)
Nicaragua 880 (1971)

Mercator, Gerhardus, *1512–1594*
Belgium 543 (1962); B324 (1942)

Michelson, Albert A., *1852–1931*
Sweden 769, 771 (1967)

Monge, Gaspard, *1746–1818*
France B279 (1953)

Morse, Samuel F. B., *1791–1872*
Monaco 611 (1965)
U.S.A. 890 (1940)

Mozhaisky, A. F., *1825–1890*
Russia 2772 (1963)

Mutis, José Celestino, *1732–1808*
Colombia 555 (1947)

Naoroji, Dadabhai, *1825–1917*
India 376 (1963)

Napier, John, *1550–1617*
Nicaragua C747 (1971)

Nasr-ud-din of Tus, *ca. 1256*
Persia (Iran) 1050 (1956)

Newton, Isaac, *1642–1727*
Ascension 142 (1971)
France 861 (1957)
Hungary 2485 (1977)
Mexico C377 (1971)
Nicaragua 877 (1971)
Niger Republic C124 (1970)
Paraguay (unlisted)
Poland 884 (1959)
Y.A.R. (unlisted)

Nicholas of Cusa. *See* Cusanus

Oersted, Hans Christian, *1777–1851*
Denmark 329 (1951); 471 (1970)

Ortelius, Abraham, *1527–1598*
Belgium B325 (1942)

Ostrogradsky, Mikhail V., *1801–1861*
Russia 1604 (1951)

Pascal, Blaise, *1623–1662*
France 1038 (1962); B181 (1944)

Perrin, Jean Baptiste, *1870–1942*
France 609 (1948)

Petr, Karel, *1868–?*
Czechoslovakia 1100 (1962)

Planck, Max K. E. L., *1858–1947*
D.D.R. 63 (1950); 383, 384 (1958)
Germany (Berlin) 9N92 (1952–53)

Planté, Gaston, *1834–1889*
France 821 (1957)

Plateau, Joseph Antoine Ferdinand, *1801–1883*
Belgium 356 (1947)

Poincaré, Jules Henri, *1854–1912*
France B270 (1952)

Popov, Alexander Stepanovich, *1859–1905*
Bulgaria 722, 723 (1951); 1126 (1960)
Czechoslovakia 703 (1955)
Hungary C62 (1948)
Romania 1267 (1959)
Russia 328, 329 (1925); 989–91 (1945); 1352–54 (1949); 1759, 1760 (1955); 2179, 2180 (1959); 3040 (1965)

Ptolemy, *ca. 127–151*
Y.A.R. (unlisted)

Pupin, Michael I., *1858–1935*
Yugloslavia 594 (1960)

Pythagoras, *ca. 540* B.C.
Greece 582–85 (1955)
Nicaragua C748 (1971)
Surinam B189 (1972)

Quételet, Lambert A. J., *1796–1874*
Belgium 877 (1974)

Quevedo, Leonardo Torres, *1852–1939*
Spain C146 (1955)

Ramanujan, Srinivasa, *1887–1920*
India 369 (1962)

Rayleigh, John William Strutt, *1842–1919*
Sweden 673, 675 (1964)

Riese, Adam, *1492–1559*
Germany 799 (1959)

Roemer, Ole, *1644–1710*
Denmark 293 (1944)

Roentgen, Wilhelm Konrad, *1845–1923*
Germany 686 (1951)
Spain 1460 (1960)
Sweden 603–5 (1961)

Russell, Bertrand, *1872–1970*
Grenada 402 (1971)
India 561 (1970)

Rutherford, Ernest, *1871–1937*
Canada 534 (1971)
New Zealand 487, 488 (1971)
Romania 2311 (1971)
Russia 3888 (1971)

Sabbah, Hassan Xamel al-, *1894–1935*
Lebanon C622 (1971)

Singh, Jai II, *1699–1744*
India-Jaipur 35 (1931)

Stevin, Simon, *1548–1620*
Belgium B321 (1942)

Struve, Friedrich von, *1793–1864*
Russia 2970 (1964–65)

Swedenborg, Emanuel, *1688–1772*
Sweden 264, 266, 267 (1938)

Teixeira, Francisco Gomes, *1851–1932*
Portugal 751, 752 (1952)

Tesla, Nikola, *1856–1943*
Czechoslovakia 949 (1959)
Yugoslavia 136, 137 (1936); 373, 374 (1953); 448–51 (1956)

Thomson, Joseph John, *1856–1940*
Sweden 710 (1966)

Titeica, Gheorghe, *1873–1939*
Romania 596 (1945); 1415 (1961)

Torricelli, Evangelista, *1608–1647*
Italy 754 (1958)
Russia 2165 (1959)

Toscanelli, Paolo dal Pozzo, *1397–1482*
Dominican Republic 105 (1899–1900)

Tsiolkovsky, Konstantin, *1857–1935*
Nicaragua 879 (1971)
Russia 1582 (1951); 1991 (1957); 2886 (1964)

Tsu Ch'ung-Chih, *429–500*
China 246 (1955)

Ulloa, Antonio de, *1716–1795*
Ecuador 348, 350, C39 (1936)

Valéry, Paul Ambroise, *1871–1945*
France B290 (1954)

Vega, Jurij (Georg), *1756–1802*
Yugoslavia 417 (1955)

Vinci, Leonardo da, *1452–1519*
Aegean Islands C8–C14 (1932)
D.D.R. 104 (1952)
France 682 (1952)
Hungary C109 (1952)
Italian Colonies C1–C7 (1932)
Italy 347, 348 (1935); 404 (1938); 601, 601B (1952); C28–C34 (1932); C103, C105 (1938)
Latvia CB10 (1932)
Liechtenstein C24 (1948)
Niger Republic C126 (1970)
Poland B73 (1952)
Romania 878 (1952)
Trieste 145, 164 (1952)

Volta, Alessandro, *1745–1827*
Italy 188–91 (1927); 527 (1949)
Somalia Italiana 97–99 (1927)

Zaviska, Frantisek, *1879–1945*
Czechoslovakia 1100 (1962)

Zeki, Salih, *1864–1954*
Turkey 1620 (1964)

Zhukovski, Nikolai, *1847–1921*
Russia 838–40 (1941); 1098, 1099 (1947); 2774 (1963)

Appendix B

A Checklist of Stamps—Applications of Mathematics to Science

The stamps listed here depict aspects of science and technology that are related to mathematics, such as mathematical symbols and mathematical instruments, calculating machines and computers, measuring devices and the metric system, surveying and cartography, space and astronomy, geometry in architecture, nuclear physics and atomic energy, and so on. Frankly, this list does not begin to exhaust the possibilities; it is far less complete than the list in Appendix A. Many more stamps could have been included under each of these headings, and many other relevant topics could have been added. This genial task is left as a challenge to the reader.

Under each category are listed the names of the countries that have issued stamps in that area of mathematical applications. The Scott number follows the country, and the date of issue is given next in parentheses. (See Appendix A for an explanation of abbreviations of countries.)

Abacus
- Australia 531 (1972)
- Surinam B188 (1972)

Arches
- Australia 429 (1967)
- Belgium B69, B70 (1928)
- Brazil 695 (1950)
- France 623 (1949); 649 (1951)
- Malta 317 (1965)
- Portugal 947–49 (1965)
- Spain 1389 (1966); 1478 (1967); 1537 (1968)
- Turkey 1796 (1968–69)

Astrolabes
- Argentina C48 (1916)
- Ascension 141 (1971)
- Persia (Iran) 1049 (1956)
- Portugal 658–61 (1945)
- Portuguese Guinea 297 (1960)
- St. Kitts-Nevis 216 (1970)
- Togo Republic 845 (1973)

Astronomy
- Albania 740–44 (1964); 777–85 (1964)
- Ascension 139 (1971)
- Finland 374 (1960)

Great Britain 616 (1970)
Japan 732 (1961)
Mexico 774–76, C123–C125 (1942)
Morocco 302 (1973)
Poland 1345 (1965)
Togo Republic 842, 844, C200 (1973)
U.N. C8 (1963–64)

Automation
Albania 1253 (1969)
Australia 440 (1968)
D.D.R. 812 (1966); 902 (1967); 1086 (1969); 1452 (1973)
Ivory Coast 329 (1972)
Russia 3081 (1965); 3416 (1967)
Tunisia 496 (1968)

Aviation
Dahomey C166 (1972)
France C183–C184 (1972)
Great Britain 584 (1969)
Russia 840 (1941); 945 (1944); 2772, 2774 (1963)
San Marino 509–18(1962)
Spain C56 (1930)
Switzerland C45 (1949)
U.S.A. C45 (1949)

Calculating machines
Denmark 415 (1965)
Germany 1123 (1973)
Iraq 390–92 (1965)

Catenary
Brazil 815 (1955)
Rhodesia and Nyasaland 173 (1960)
Russia 2009 (1957)
Switzerland 328 (1949)
Vietnam (South) 227–30 (1964)

Circular arches (bridges)
Czechoslovakia 727 (1955)
France 253, 254 (1929–33); 623 (1949); 831 (1957); 982 (1960)
Great Britain 561 (1968)
Hungary 1143 (1955)

Compass (mariner's)
Algeria 351 (1966)
Libya 276–78 (1965)
Mexico C313 (1966)
Vatican City 424 (1966)

Compasses (dividers)
Albania 794 (1965)
Argentina C56 (1948–49)
Brazil 729 (1952)
Finland 397 (1962)
Papua and New Guinea 234 (1967)
Zanzibar 349–51 (1966)

Computation
Brazil 739 (1953)
Iran 1413 (1966)
Nicaragua 876 (1971)
U.A.R. (Egypt) 385 (1969)

Electricity
D.D.R. 533 (1961)
Denmark 471 (1970)
Germany 965, 966 (1966); 1005 (1969)
Nicaragua 880 (1971)
Spain 1460 (1960)
U.S.A. 1073 (1956); 1500 (1973)
Yugoslavia 448–50 (1956)

Electronic computers
Australia 531 (1972)
Canada 542 (1971)
D.D.R. 811–12 (1966); 1532 (1974)
Israel 258 (1964)
Netherlands 451 (1968); 487 (1971)
Norway 547 (1968)

Electronic computers (uses of)
Albania 1253 (1969)
Australia 440 (1968)
D.D.R. 812 (1966); 902 (1967); 1086 (1969); 1452 (1973)
Ivory Coast 329 (1972)
Russia 3081 (1965); 3416 (1967)
Tunisia 496 (1968)

Elliptic arches (bridges)
Chile 390 (1970)
France 255 (1929); 421 (1941); 498 (1944)
Lebanon 238–42 (1950)
Paraguay C226 (1955)

Geodesy
Ecuador 347–51, C39–C42 (1936)
Finland 373 (1960)
U.S.A. 1088 (1957)

Geodesic dome
Hungary 1589 (1964)
Ras al Khaima (unlisted)
Russia 4314 (1975)

Globe (earth)
Brazil 873 (1958)
D.D.R. 1402–7 (1972)
Italy 718, 719 (1956)

Graphs (bar)
Korea 430 (1964)
U.N. 137, 138 (1965)
Venezuela 438–44, C302–C310 (1950)

Graphs (broken line)
Bolivia 472 (1963)
Cuba (unlisted)
Denmark 412 (1964)
Iraq 390–92 (1965)
Israel 218 (1962)
Kuwait 275–77 (1966)
Poland B59–B62 (1918)
Romania 727, 728, 756 (1950); 827C, 844A (1952)
Spain 853–55 (1956)
Turkey B79 (1960); B93, B94 (1963)
U.A.R. 97 (1961)

Graphs (census charts)
Brazil 697 (1950)
Chile 277–80 (1953)
Costa Rica C403 (1965)

Graphs (pictorial)
Korea 317 (1960)
Turkey 1182–85 (1955); B80 (1960)
U.N. 151–53 (1965); 185 (1968)

Map projections
Australia 283 (1955)
Canada 85, 86 (1898); 411 (1963)
Chile 247, 248 (1917); 305 (1958); C214 (1959)
France B115 (1941)
Japan 1180, 1181 (1974)
Liberia 379 (1958)
Mexico C271 (1963)
U.N. 125 (1964)
U.S.A. 1112 (1958); 1274 (1965)

Mathematical symbols
Canada 396 (1962): $E = mc^2$
Colombia 594 (1951); 742, C132 (1962): π
Columbia C510 (1968): $+, -, \div, \times$
D.D.R. 383 (1958): h (quantum)
Germany 1246 (1977): Complex numbers
Poland 1178 (1968): $v = w \ln m_0/m$
Romania 1876 (1966): integral
Russia 3151 (1966); 4464 (1976): integral
Rwanda 84, 88 (1965): $\sqrt{}$
Switzerland B303 (1961): ∞

Measuring instruments
India 392 (1964)
Japan 673 (1959)
Korea 803 (1971)
Romania 1159, 1160 (1957)
U.S.A. 1201 (1962)

Metric system
Australia 541–43 (1973)
Brazil 940 (1962)
France 732 (1954); 1435 (1975)
Korea 428 (1964)
Mexico C241 (1957)
Pakistan 364 (1974)
Romania 1873, 1874 (1966)
Tanzania 225–28 (1971)
Yugoslavia 1180 (1974)

Nuclear power
Argentina C116 (1969)
Belgium 537 (1961)
China 392 (1958)
D.D.R. 465 (1959)
France 921 (1959)
Greece 716 (1961)
Hungary 1763 (1966)
Israel 182 (1962)
Japan 638 (1957)
Korea 349 (1962)
Pakistan 187 (1963)

Romania 1360 (1960)
Russia 1794–96 (1956)
Vietnam (South) 231–34 (1964)
Yugoslavia 582–84 (1960)

Observatories
Argentina 969 (1971)
China 1337 (1962)
Haiti 426 (1958)
Japan 478 (1949); 591 (1953); 707 (1960)
Korea 69 (1946); 94 (1948)
Russia 1722 (1954); 1958 (1957); 2092 (1958)
Turkey B81, B84 (1961)
U.S.A. 966 (1948)

Parabolic antenna (radar)
Argentina C115 (1969)
Chile 402 (1971)
Jordan 856, 857 (1975)
Mozambique 510 (1974)
Switzerland 555 (1973)

Parabolic arches (bridges)
Brazil 398, 401 (1935)
China C73 (1963)
El Salvador C66–C68 (1939); C92 (1944); C159 (1954)
France 683 (1952); 926 (1959)
Ghana 293 (1967)
Great Britain 418, 419 (1964)
Hungary C64, C65 (1949)
Japan 768 (1962)
Portugal 757 (1952)
Russia 1132 (1947)
Switzerland 311 (1947)
U.S.A. 961 (1948); 1109 (1958); 1258 (1964)

Physics (atomic)
Argentina 951 (1971)
D.D.R. 383 (1958)
Denmark 409, 410 (1963)
Greenland 57, 58 (1963)
Nicaragua 878, C750 (1971)
Russia 2721 (1963); 3888 (1971)
Sweden 790 (1968)
U.N. 59, 60 (1958)
U.S.A. 1500–1502, C86 (1973)

Physics (classical)
D.D.R. B150 (1967)
Germany 892–94 (1964); 980 (1968)
Israel 256 (1964)
Nicaragua 877, 880, C749, C751 (1971)
Pakistan 281 (1961)
Spain 1487 (1967); 1570 (1969)

Pythagorean theorem
Greece 583 (1955)
Nicaragua 748 (1971)
Surinam B189 (1972)

Quadrant
Ascension 140 (1971)
Turkey B82 (1961)

Radio
Austria 739 (1964)
Czechoslovakia 954 (1959); 1176, 1177 (1963)
D.D.R. 769, 770 (1965); 1204, 1205 (1970)
Germany (Berlin) 9N343 (1973)
Russia 3526 (1968)
U.S.A. 1329 (1967); 1500–1502, C86 (1973)

Radio waves
Argentina 922 (1970)
Central African Republic C29 (1965)
Czechoslovakia 1529–30 (1968)
Germany 932 (1965)
Israel 256, 257 (1964)
Norway 471 (1965)
Portugal 1181 (1973)
U.S.A. 1260 (1964); 1274 (1965)

Rocketry
Bulgaria C92 (1962)
France 1184 (1967)
Nicaragua 879 (1971)
Poland 1178 (1963)
Russia 1991 (1957)
U.S.A. C69 (1964)

Sextant
Australian Antarctic Territory L21 (1972)
New Zealand 431 (1969)
Nicaragua 747 (1971)

Slide rule
Romania 1159, 1160 (1957)

Space satellites (and telecommunication)
Ascension 104–7 (1966)
Canada 444 (1965)
Chile C290 (1969)
Colombia C499 (1968)
Congo Republic C27 (1965)
D.D.R. 370 (1957)
France 1148 (1968)
Malagasy 465 (1972)
Nigeria 143 (1963)
Poland 875 (1959)
Romania C49–C51 (1957)

Surveying
D.D.R. 726 (1964)
Finland 373, 374 (1960); 397 (1962)
Mexico 923 (1962)
Nigeria 207 (1967)
Papua and New Guinea 233 (1967)
Qatar 324 (1973)
Russia 2554 (1961); 2656 (1962)
Spain 1635 (1970)
Switzerland 339 (1949)
Tanzania 228 (1971)

Telescopes (optical)
Ascension 141, 142, 145 (1971)
Czechoslovakia 1231 (1964)
D.D.R. 898 (1967)
Japan 478 (1949)
Russia 1957 (1957); 2094 (1958)

Telescopes (radio)
Ascension 146 (1971)
Czechoslovakia 836 (1957)
France 1067 (1963)
Great Britain 466 (1966)
Haiti C121 (1958)
Israel 496 (1972)
Japan 799 (1963)
Sharjah and dependencies 88 (1965)

Index

Abacists, 35
Abacus, 25
Abel, Niels, 76
Alembert, Jean d', 71, 72
al-Farabi. *See* Farabi
Algebra, Babylonian, 7
Algorists, 35
al-Haitham. *See* Alhazen
Alhazen, 30
Alidade, 43
al-Kindi. *See* Kindi
Alouette II, 131
Ampère, André, 90
Analog computer, 120
Analysis, mathematical, 67–73
Analytic geometry, 61
  trigonometry, 69
Applications of computers, 122–25
Arabian mathematics, 29–32
Arago, François, 90
Archimedes, 21–23
Aristotle, 20, 39
Arithmetic
  Egyptian, 9, 10
  in the Middle Ages, 33–35
Armillaries, 43
Armstrong, Edwin, 101–3
Armstrong, Neil, 133
Astrolabe, 43
Astronomical instruments, 43, 45, 52–55, 130
Astronomy
  Babylonian, 7
  Greek, 23, 24, 40, 45, 46
Athenian school, 18–20
Atomic Energy Act, 118
Atomic Energy Commission, 118
Atomic structure, 111–13
  theory, 19, 20, 107–13
Atomism, 19, 20, 111
Atoms for Peace, 118
Automation, 120–25
Aviation, 104–7
Avicenna, 31, 32

Babbage, Charles, 88
Babbage's calculating machine, 88
Babylonian mathematics, 4–7, 12, 13
Babylonian numeration, 5, 6
Bacon, Francis, 55
Bacon, Roger, 28, 40
Bede, Venerable, 40
Bell, Alexander Graham, 97
Blackbody radiation, 91, 108
Bohr, Niels, 112, 113, 115, 117
Boltzmann, Ludwig, 91
Bolyai, János, 77, 78
Bolyai, Wolfgang F., 77
Boole, George, 20, 94
Boscovich, Ruggiero, 111, 112
Brahe, Tycho, 48
Branly, Edouard, 100
Broglie, Louis de, 116

Calculating machines, 62, 67, 87, 88
Calculus, 65–67, 70–72
Chaplygin, Sergei, 107

Chebyshëv, P. L., 81
Communication satellites, 130–32
Complex numbers, 69
Computation, 86–88
Concorde, 107
Copernicus, Nicolaus, 46–48
Corpuscular theory of light, 60, 64
Coulomb, C. A. de, 88, 89
Counter reckoning, 34
Cuneiform writing, 5, 6
Cusanus, 33, 34

d'Alembert, Jean. *See* Alembert
Data processing, 122–25
da Vinci, Leonardo. *See* Vinci
de Broglie. *See* Broglie
Decimal fractions, 57
De Forest, Lee, 101
Democritus, 19, 20
Descartes, René, 60–62
Digital computer, 120
Dioptra, 43
Disquisitiones Arithmeticae, 75
Dürer, Albrecht, 35–37

Egyptian mathematics, 8–13
Egyptian numeration, 9
Einstein, Albert, 109, 115, 116, 117
Einstein's letter to President Franklin D. Roosevelt, 115
Electricity, 88–92
Electronic computers, 120–25
Electronic theory, 98–103
ENIAC, 122
Eötvös, Loránd, 96
Esnault-Pelterie, Robert, 126, 127
Euclid, 20, 21
Euler, Leonhard, 67
Explorer I, 129

Farabi, al-, 29, 30
Fermat, Pierre de, 63
Fermi, Enrico, 115, 116
Figurate numbers, 17
Finger reckoning, 10
Foucault, Jean, 81, 82
Foucault pendulum, 82
Fractions
  decimal, 57
  Egyptian, 10
Franklin, Benjamin, 88, 89

Gagarin, Yuri, 132
Galilei, G., 49–52
Galileo. *See* Galilei
Gauss, C. F., 74–79
Geodetic surveying, 84
Geometry
  analytic, 61
  Babylonian, 6, 7
  Egyptian, 10–12
  Greek, 17–23
  non-Euclidean, 77–79
Gerbert, 32, 33
Glenn, John, 133
Goddard, Robert H., 126, 127
Gravitation, law of, 65
Great Pyramids (of Egypt), 11, 12
Greek culture, 14, 15, 26
Guericke, Otto von, 58, 59
Gutenberg, Johann, 28, 29

Hahn, Otto, 115
Haitham, al-. *See* Alhazen
Hamilton, William Rowan, 79, 80
Hell, Maximilian, 53
Herschel, William, 72
Hertz, Heinrich, 92
Hertzian waves, 92, 98
Hindu-Arabic numerals, 29, 30, 34
Hipparchus, 23, 24
Hiroshima, 117
Huygens, Christian, 59, 60

ibn-al-Haytham. *See* Alhazen
Imaginary numbers, 69
Imhotep, 9
Industrial uses of computers, 122–25
Ionic capital, 15
Instruments
  astronomical, 43, 45, 52–55, 130
  calculating machines, 62, 67, 87, 88
  computational, 86, 87
  measuring, 82–85
  surveying, 83–85
INTELSAT systems, 130

Kepler, Johannes, 49
Kepler's laws, 49
Khayyám, Omar, 32
Khwarizmi, al-, 30
Kindi, Abu-Usuf al-, 29, 30
Kirchhoff, Gustav, 90, 91
Königsberg bridges, 69, 70
Kovalevski, Sonya, 80
Kremer, Gerhard, 40, 41

Lagrange, J. L., 70, 71
Langevin, Paul, 96
Laplace, P. Simon de, 72

Law(s)
of Archimedes (lever), 22
of Coulomb, 89
of Einstein, 110, 116
of falling bodies, 50
of gravitation, 65
of Kepler, 49
of lever, 22
of Maxwell, 92
of motion, 65
of planetary motion, 49
Stefan-Boltzmann, 91
Tsiolkovsky, 126, 128
Leibniz, G. W. von, 65–67
Leibniz' calculating machine, 67
Leonardo da Vinci. *See* Vinci
Light
corpuscular theory of, 60, 64
velocity of, 81, 82, 109, 110
wave theory of, 60, 64
Lindbergh, Charles, 104
Lobachevski, Nikolai, 78
Lorentz, Hendrik, 107, 108

Magdeburg hemispheres, 58, 59
Maimonides, 32
Manned space flights, 132–34
Map projections, 41–43
Marconi, Guglielmo, 98–100
Mathematical analysis, 67–73
Mathematical models, 136
Mathematics
in the Middle Ages, 27–29, 38
relationship of, to science, 135, 136
structure of, 79, 94, 136
Maxwell, James Clerk, 91, 92
McMahon, Brien, 118
McNamara, Robert, 117
Measurement, 82–86
Measuring instruments, 82–85
Mécanique analytique, 70
Mécanique céleste, 72
Mercator, Gerhardus, 40, 41
Mercator map projection, 40, 41
Mercury (U.S.) capsule, 130
Metric system, 71, 85, 86
Michelson-Morley experiment, 109
Micrometers, 83
Moon landing, 134

NASA, 132
Newton, Isaac, 63–66
Newtonian physics, 110
Nicholas of Cusa. *See* Cusanus
non-Euclidean geometry, 77–79
Nuclear age, 115
Nuclear physics, 110
Nuclear power, 118, 119
Nuclear warfare, 117, 118
Number mysticism, 16
Numeration
Babylonian, 5, 6
Egyptian, 9
Roman, 25

Observatories, 53
Oersted, H. C., 89, 90
Optical telescope, 52, 53
Optics, 30
Ortelius, Abraham, 40, 41

Parallel sailing, 44
Pascal, Blaise, 62, 63
Pascal's calculating machine, 62
Pendulum, 50, 60, 82
Pendulum clock, 59, 60
Pentagram, 17
Perrin, Jean, 96
Perspective (in painting), 36, 37
Planck, Max, 108, 109
Planck's constant, 108
Planetary motion, 49, 65
Planisphere, 43
Poincaré, Henri, 94, 95, 96
Pope Sylvester II, 32, 33
Popov, Alexander, 100–102
Probability, 63
Ptolemy, 24, 40, 45, 46
Pyramids (of Egypt), 11, 12
Pythagoras, 16–18
Pythagorean theorem, 17, 18
Pupin, Michael, 97

Quadrant, 45
Quantum mechanics, 108
Quaternions, 79, 80
Quételet, L. A. J., 81

Radio, 101, 102
Radio telescope, 53, 54, 130
Radioactivity, 98
Ramanujan, S., 95, 96
Relativity theory, 107–13
Rhind papyrus, 8, 9
Riese, Adam, 34
Rocketry, 125–28
Roentgen, Wilhelm, 97
Roman mathematics, 24, 25
Roman numerals, 25
Rosetta stone, 8
Royal Astronomical Society, 73
Russell, Bertrand, 95, 96
Rutherford, Ernest, 98, 112

Satellites, 128–32
Scientific method, 28, 40, 55
Sextant, 45
Simulation method, 122
Skylab, 133
Slide rule, 87
Space astronomy, 130–32
Space pioneers, 133
Sputnik I, 128, 129
Stefan-Boltzmann law, 91
Stevin, Simon, 57, 58
Structure
  of the atom, 111–13
  of mathematics, 79, 94, 136
Surveying, 83–85
  geodetic, 84
Swann, W. F. G., 136
Symbolic logic, 20, 94
Szilard, Leo, 115, 117

Telecommunication, 130–32
Telescope
  optical, 52, 53
  radio, 53, 54, 130
Television, 102–4
TELSTAR, 130
Tesla, Nikola, 98, 99
Thales, 16
Theodolite, 84
Torricelli, Evangelista, 58
Transit level, 84
Triangulation, 84
Trigonometry, 23, 24
  analytic, 69
Tsiolkovsky, Konstantin, 126–28

Vanguard I, 129
Vinci, Leonardo da, 35–36, 104
Volta, Alessandro, 89
von Neumann, John, 122

Wave theory of light, 60, 64
Weinberg, Alvin, 119
Wireless communication, 92, 98–101
Wright, Orville and Wilbur, 104, 105

Zhukovski, N. E., 107